T0133901

Information and Communication Technology for Sustainable Development

Information and Communication Technology for Sustainable Development

Cesar Marolla

CRC Press
Taylor & Francis Group
Boca Raton London New York

CRC Press is an imprint of the
Taylor & Francis Group, an **informa** business

CRC Press
Taylor & Francis Group
6000 Broken Sound Parkway NW, Suite 300
Boca Raton, FL 33487-2742

Visit the Taylor & Francis Web site at
http://www.taylorandfrancis.com

and the CRC Press Web site at
http://www.crcpress.com

To my wife Lisa for standing beside me throughout my career and writing this book.

She is my rock, and I dedicate this book to her. I also thank my wonderful children:

Nicholas, Alexander, and Michael, for giving me the courage and determination

to explore higher levels of accomplishments in my life. They are my inspiration.

To my parents, Sofia and Jose Cesar Marolla, for their

support, strength, and understanding.

Inspire, Lead and Empower Change.

– Cesar

Contents

Foreword

It's a no-brainer that information and communication technology (ICT) has enormous potential to accelerate our achievement of Agenda 2030. This is recognized by both the ICT community and the United Nations as the key proponent of the sustainable development goals (SDGs). What we need to understand is why this is so, what the risks and constraints are, and how they can be overcome.

Many ICT companies have identified the business opportunity that exists in applying technology in order to make progress with realizing SDGs. The 2017 Huawei ICT Sustainable Development Goals Benchmark explored the relationship between ICT and six SDGs as well as country performances on SDGs and ICT development. The study found:

- Countries that perform well on ICT do equally well on their SDGs, while those underperforming on ICT are also lagging on SDG achievement.
- The highest correlations between ICT development and SDG achievement are for SDG 3 (good health and well-being), SDG 4 (quality education), and SDG 9 (infrastructure).
- Developed countries generally score higher on ICT development than SDG achievement, suggesting that the former is progressing more quickly than the latter.
- The inverse tends to be the case in developing countries, where ICT enhancement is not keeping pace with progress on the SDGs.

Thus, there seems to be a particular investment opportunity in developing nations to improve ICT infrastructure, which, in turn, would have a positive feedback for particular SDGs.

The United Nations also recognizes the potential for the digital revolution to spur human progress through SDG achievement. In the UN's 2017 report on "Fast-forward progress: Leveraging tech to achieve the Global Goals," UN Secretary General Antonio Guterres notes that:

- There are risks and opportunities in using ICT to achieve the SDGs.
- Various technologies (e.g., big data, remote sensing, population mapping) can help with policy design, decision-making, environmental management, and social inclusion.
- UN agencies are adapting ICTs to help communities and people in need through, *inter alia*, digital inclusion, digital skill development and smarter schools, health care, and cities.

His conclusion is that the UN is committed to making the most of cutting-edge technologies and the opportunity that they represent to improve service delivery.

So, if the ICT and SDG communities are on the same wavelength, what are the opportunities and challenges? The World Bank's 2016 *World Development Report* on digital dividends can provide us with some insights. First, ICTs have benefited people, firms, and societies by expanding access to information and services, lowering the costs of economic and social transactions, and creating information goods that foster innovation. However, the development impacts of ICT have fallen short of their potential because the digital divide is still large (nearly 60% of the global population is still offline) and gains from the digital revolution have not been widely shared. The emerging development risks from ICT include: (a) amplification of the voice of elites and their social and economic power; (b) a hollowing out of the labor market through automation; and (c) market concentration where the internet favors natural monopolies and a less competitive business environment. And there are technological risks such as content filtering, censorship, cybercrime, invasion of online privacy, and manipulation of information ("fake news").

Given these opportunities and challenges, what can be done to enhance the contribution that ICTs make to more sustainable development? The *WDR* concludes that broad-based gains will be elusive if technological innovation is not matched with improvements in other factors. These "analog complements" include a country's business environment, the adaptability of education, social protections and labor markets, and public sector accountability (whether digital technologies are used to control or empower citizens). Thus, the way forward lies not just in technological innovation for particular sustainable development goals, but also requires progress with regulations that allow firms to connect and compete, building human capital and skills, and developing institutions with capacity and accountability. This volume provides us with useful guidance in these areas so that digital dividends can help achieve a more sustainable world with higher-quality economic growth, shared prosperity, and greater resilience.

Dr. Joe Leitmann
Team Leader,
Resilient Recovery and Urban Resilience –
Global Facility for Disaster Reduction and
Recovery/World Bank.

Preface

Sustainable development entails the action of making effective decisions from some set of available alternatives for the best use of a situation or resource. Human societies have made important technological achievements in the past two centuries. However, besides the digitalization of business and society and technological advancement in recent years that has a major impact on each of us individually, our present and future survival is intrinsically dependent upon the continued functioning of natural systems, which makes the intertwined relation between the environment and technology crucial for our own existence. The standard of general well-being in health, comfort, work, and happiness experienced by individuals and society as a whole is commensurable to the values which we can influence in society. Sustainable development (SD) faces many challenges in urbanization, sustainability, energy efficiency, eradicating poverty, etc.; however, information and communication technology (ICT) is emerging at the beginning of the twenty-first century as an important field to have the ability to search, locate, evaluate, manage, use, present, and communicate information and to engage in problem-solving, self-directed learning, autonomous-targeted learning, and research skills, which are all fundamental to reach sustainable development goals (SDGs). ICT is seeking to improve environmental effectiveness in the context of connected communities, global competitiveness, economic development, climate change, and demographic shifts. Virtually, all proposed solutions to energy consumption and climate change acknowledge the role ICT plays as a key enabler of environmental effectiveness (Markovic et al., 2012).

We live in a rapidly changing world where we face some of the greatest challenges of our time and ICTs are playing an increasing role in enhancing local economic development, poverty reduction, and the overall well-being of our society. From the local to the global level, ICTs are tackling all areas of risks, addressing a growing series of short-term shocks (economic crises, severe weather events, violent attacks, health epidemics, etc.) and long-term trends (climate change, migration, economic restructuring, new health care technologies, etc.). Leaders around the world are embracing the capacity of ICT to mitigate the harmful effects of environmental degradation, air pollution, natural disasters, national security, and other risks. Nevertheless, the public health sector is placed in an optimum position to impose a responsibility for policy-makers, and indeed all stakeholders of the information society (IS), to promote the technology as an effective tool to combat global disease detection, response, prevention, and control strategies and assessing serious public health threats with detrimental implications to public health. Furthermore, the link between ICT and strategy is more important now than ever before. The strategic approach aligned with technology is essential to enable strategic objectives and achieve sustainable development goals.

The book explores how the adoption of new technologies and collective action based upon a strategic behavioral theory of new leadership responding to specific set of conditions that allows for the proposed strategies to cope with risks, can be applied. Moreover, it addresses the interlink between climatic impacts affecting the public and private sector (urban and rural populations, organizations, cities, etc.) and the role of ICTs to deter and/or minimize the effects of climate variability and change on public health, and address underdevelopment as a social and economic phenomenon. As such, it appears important to develop strategic frameworks within collective actions made by leaders of the public and private sector that consequently becomes achievable; where success is no longer solely tied to individual incentives, but equally to the provision of information, learning, and interaction between local and global stakeholders, while simultaneously fostering ICTs as the cross-cutting technology that can drive the deep transformation needed in the global effort to create a sustainable world.

ICTs, along with the appropriate strategic planning in place, can enhance environmental performance, support existing telecommunication systems and health care, improve illness prevention and safety of care, facilitate active participation of patients, and enhance continuity management systems. This will minimize the impact of disruptions and address the integration of the economic, social, and environmental dimensions in policy-making at international, regional, and national levels. The opportunities across the different networks within a city, organization, or community are in power generation and distribution, buildings, and transportation and these areas contribute to the highest degree to greenhouse gas (GHG) emissions. Further, environmental benefits of ICT applications are evident in areas such as water management, biodiversity protection, and pollution reduction. Hence, the tools and benefits of ICTs are essential to spur growth and improve the health of the most poor and vulnerable people. Most government and business initiatives are still following business-as-usual paths and do not sufficiently tap the sector's innovation potential. This is because most nations do not have efficient economic and environmental strategic models to build on existing assets, create new economic and environmental models, and maximize the return on investment while taking concrete actions to protect the environment and human health.

The essential role of ICT sits in monitoring, modeling, management, and dissemination. As global risks such as climate change and cyberterrorism, which are widely considered as two of the most significant threats of our time, become more complex and predominant, these courses of action need to be effective. The role of ICTs becomes crucial to adaptation and mitigation strategies to manage and consequently deter those impacts. Some examples are monitoring weather and climate variability and change, and deforestation using satellite imagery; environmental modeling such as computer simulations; administrative processes using emissions/carbon trading schemes for trading greenhouse gas emissions allowances; and dissemination, including information sharing and environmental influence in public policy within social, economic, political, and cultural spheres (Measuring the Relationship between ICT and the Environment, 2009). Nevertheless, cyberattacks are a threat to national security and have different approaches and forms such as a massive distributed denial of service (DDoS) attack to bring down parts of a national or international information infrastructure. That is the reason to plan, implement, and upgrade security measures and controls and implement risk management frameworks and business continuity systems to manage network, intrusion detection, and prevention systems.

Another angle of the equation to solve sustainable development challenges is in the health sector. Holmner et al. (2012) stated that the health sector in developed countries contributes significantly to greenhouse gas emissions (GHGs) and in developing countries its impacts are increasingly growing. The health sector prospective for climate change adaptation and mitigation strategies are playing an important role to manage its detrimental effects. The health information technology such as eHealth has potential to reduce emission rates from ICT use and build up adaptation strategies to lessen societal vulnerability to climate change. As local and international initiatives are in place to combat climate change, reduce emissions, and develop adaptation measures, ICTs are now mentioning eHealth as a promising technology to mitigate GHGs and adaptation policies.

At the United Nations Sustainable Development Summit on September 25, 2015, world leaders adopted the 2030 Agenda for Sustainable Development, which includes a set of 17 Sustainable Development Goals (SDGs) to end poverty, addressing responsible consumption and production, sustainable cities and communities, fight inequality and injustice, and tackle climate change by 2030 (Sustainable development goals, 2018).

The SDGs have specific measures comprising 169 targets and they are measured using one or more indicators to assess progress in an efficient way. This thorough set of objectives was designed to build upon the work done in the context of the Millennium Development Goals (MDGs) and aimed at specific socio-economic situations adapted to the different realities of the developed and developing countries, placing greater importance on addressing environmental concerns. Our common pursuit of growth, poverty eradication, and access to health care and sustainable development should incentivize analytical thinking and strategic leadership, ensuring the necessary internal and external conditions for mobilizing the private and public sector, sustaining adequate levels of productive investment, and increasing human capacity to find tangible solutions. In regard to climate change impacts, which also affect our health, denoting shifts in conditions (such as sea level rise, melting glaciers, or changing oceanic acidity due to atmospheric CO_2 uptake) which happen over long periods of time, ICTs can play an immeasurable role in mitigating, adapting to, and monitoring those effects. As more events like record-setting heat, drought, and extreme rainfall keep occurring, in addition to scarcity of natural resources, migration, and other implications of climate change effects, leaders need to set research questions, objectives, causes and contributions of the ICT sector to climate change. Nevertheless, adaptation and mitigation with an all-inclusive risk management and continuity planning must be tailored to the strategy. Without a clear strategic direction, any initiatives and actions toward the goals are doomed to fail. The world needs leaders with a deep understanding of technology and management. The leader's role in promoting technology integration and minimize risks is essential to a wide-ranging strategy. The book defines the challenges and opportunities for realizing the promises of ICT-enabled development in terms of the institutional transformations and change processes that must be led, inspired, and coordinated. This context provides the basis for defining the demand for e-leaders and the core competencies required of them and of e-leadership institutions. We need to emphasize the special importance of timely, effective, comprehensive, and long-term solutions. I am a strong proponent of improving the rigor in sustainability initiatives and intelligence analysis, and I am a believer for better grounding of new analysis in formal logic, cognitive theory, and structured techniques of creative problem-solving. At the same time, however, I am deeply mindful of the continuing unmet needs for more creative and imaginative analysis, which moves further afield from the narrowly constructed questions of current global uncertainties, particularly regarding climate variability and change, natural disasters, public health, and the spread of diseases and national security implications of cyber-attacks leading into the realm of the "unknown unknowns." In that context, strategy and sustainability are intrinsically connected. As Michael Porter, professor at Harvard Business School expressed: "Society's challenges represent the next wave of innovation and productivity growth in the global economy." Recognizing the close links that exist between ICTs and the achievement of sustainable development goals (SDGs), alongside the rapid increase in investments in ICT infrastructure and projects, the book put at front a conceptual foundation that links global risks and challenges such as poverty eradication, public health risks, climate change, smart cities, water scarcity, and energy consumption just to name a few, and the potential of ICTs in supporting systemic resilience toward the aforementioned issues. ICTs are introduced as a system component that has the potential of contributing toward resilience and, therefore, helping to enable strategies and unchain a world that is made up of networks. The development of an integrated quantitative measure and modeling of sustainable development and disaster risk reduction is needed and resilience frameworks, which has been used to create partnerships and communicate pre- and post-disaster recovery of extreme events and ICTs, can partner to achieve a systemic perspective. This allows the identification of key components, processes and properties, as well as the feedback and interactions that play a role in the realization of adapting processes in vulnerable settings. Within the emerging field of

ICTs and its role in combating climate change and environmental and health risks, and promoting economic growth and the leader's role in strategy, the book responds to the need for building a solid conceptual basis upon which to analyze the function and potential of these tools, while recognizing existing development challenges and vulnerabilities. ICTs have the prospects to create a new, inclusive, social, and economic order and for leaders to take actions that could be sustained without undermining future generations, finding solutions to our biggest environmental and societal challenges and transforming individual lives, as well as local and national economies for a transformative and sustainable world.

Nonetheless, it is important to have a sustainable development strategy. Setting targets, developing plans, and other means are necessary to establish a coherent foundation in achieving sustainable development goals. For sustainable development to materialize in every aspect of the nation's functions they need all-inclusive management frameworks and strategic approaches that are "universal." In this context, universal means strategies to cope with the economic, social, and environmental challenges they face and address its obstacles with "common denominators" or proven strategic planning that can create socially desirable outcomes. There are significant efforts made by many nations to achieve SDGs, but overall those efforts are static and isolate each nation and its leaders, generating confusion with no clear outcomes; working in silos leading to some accomplishments but not to a thorough understanding of the issues they face or, more importantly, how to find pragmatic solutions to achieve SDGs.

I propose, along with ICT's role in accelerating sustainable development, comprehensive management frameworks and strategies to cope with those risks that put in danger the viability of sustainable development, including the vital participation of many agents of the public and private sector working together for common goals. Many areas of sustainable development remain, in principle, unclear, making it difficult to plan an effective course of action to accomplish the desired goals. Information technology and strategic planning complement each other to attain the aforesaid SDGs. Risk management frameworks, business continuity systems, and strategic methodologies such as mechanism design theory, strategic adaptive cognition (SAC), and risk mechanism theory (RMT) are the fundamental components needed to have a universal approach embedded into the national development plans agenda and help world leaders in the difficult but inspiring task of making a sustainable world for the well-being of the present and future generations.

Cesar Marolla

Acknowledgments

There are many barriers to tackling global risks that avert sustainable development. It is essential to work together within all sectors of society, the public and private segments, and the academic, scientific, environmental, and government communities in the creation of a new global decision-making body to manage threats to humanity. I thought about writing this book for many years simply because I witnessed a wide gap between technology, sustainability, and strategic leadership while working in the information and communication technology field. Nevertheless, my speaking engagements with world leaders at the United Nations Conference of the Parties, Harvard, and the University of Miami, just to mention a few, made me realize the need to create a coherent, intertwined strategy that integrates the scientific community findings, management, and strategic leadership approach with economists, technology and sustainability experts to find common grounds into the development of a concise methodology to achieve sustainable development.

Subsequently, my conversation regarding these important issues with Harvard Business School Professor John P. Kotter opened my world to new possibilities in the areas of business strategy and sustainable development. I can argue nations, businesses, cities, and other public organizations are falling short in achieving sustainable development goals, but what is even more significant is the absence of pragmatic strategies and exemplary leadership to cope with the challenges, obstacles, and critical problems the world faces today. Learning from Professor Kotter's pioneering work allowed me to visualize another dimension as a strategist and his world-renowned, highly regarded contributions support a structure for innovation and creativity that must be taken into account to solve the global risks the world faces today.

Dr. Eric Maskin, the 2007 Nobel Memorial Prize in Economic Sciences laureate, whose fundamental contributions to mechanism design and work in economic theory has had a deep influence on many areas of economics, political science, and law, made a significant influential direction in my own work developing strategies and sustainable management frameworks. Mechanism design significantly contributes to sustainable development and becomes an imperative call to developing sustainable competitive advantages in order to achieve a sustainable society. I am humbled and grateful Professor Maskin has shared his important work with me, and holds promise for more sustainable forms of economic process, dealing with such urgent issues as public health risks, carbon emissions, and resource depletion, leading to important implications for the ability of markets to deliver sustainable development.

The teachings in risk management and corporate sustainability of my former Harvard professor in sustainability and environmental management, Dr. Robert Pojasek, built my cognitive ability to break down complex actions into manageable units and prioritize them in the right order. Professor Pojasek's mentorship has been vital to the development and implementation of my research, strategic thinking, and goal-oriented behavior.

Finally, yet importantly, Dr. Joe Leitmann, Lead Disaster Risk Management Specialist at the World Bank, heading teams on Resilient Recovery and Urban Resilience at the Global Facility for Disaster Reduction and Recovery (GFDRR), is an exemplary leader who showed me through his work the meaning of impact, purpose, and inspiration. Dr. Leitmann's work has an influence and impact worldwide, changing people's lives and making concrete actions toward building the foundations of sustainable development in order to understand the true meaning of sustainability.

Therefore, I'd like to express my sincerest appreciation to each of the aforesaid world leaders for their long-lasting contributions to society and I look forward to continuing working to support tangible actions and finding solutions to the world's challenges and to ensure that the resources of this planet are utilized efficiently and sustainably.

A special note to express my gratitude to Irma Shagla Britton, Senior Editor, Environmental Sciences, GIS & Remote Sensing CRC Press/Taylor & Francis Group, for her invaluable support and encouragement throughout the challenging process of putting my work and ideas together.

Author

Cesar Marolla is a sustainability and environmental management leader, researcher, author, and lecturer. Throughout his career he has brought strategy concepts to bear on many of the most challenging problems facing corporations and societies, including working with global organizations, such as Deutsche Telekom and nongovernmental organizations (NGOs) as well as volunteering with the US Department of Defense. He has traveled extensively and worked in sustainability, climate change, and risk management, business marketing strategies, and corporate responsibility in Europe, South America, Middle East, Northeast Africa, and the United States. Marolla has assembled sustainability assessments and best practices for Fortune 500 corporations and participated in climate talks with international organizations dedicated to communicating how the information and communication technology (ICT) industries can address and provide solutions for environmental issues. He has interviewed, researched, and collaborated with world-renowned leaders from the World Health Organization (WHO), Deutsche Telekom, US Centers for Disease Control and Prevention (CDC), United Nations, World Bank, University of Miami, Columbia University, and Harvard University, in addition to city mayors in issues such as climate change adaptation and mitigation strategies, environmental management, risk management, and public health risks.

Cesar is the recipient of the 2013 Harvard University Derek Bok Civic Prize Award that recognizes creative initiatives in community service and long-standing records of civic achievement. He also received a Certificate of Appreciation from the Pentagon for his volunteerism supporting US troops in the Middle East and Northeast Africa under the umbrella of the Global War on Terror (GWOT). Moreover, he received a "Military Coin," which is given as a token of affiliation, support, patronage, respect, honor, and gratitude, and presented by the Camp Victory Commander in Kuwait, Lieutenant Colonel Lawrence J. Smith. Cesar earned a bachelor's of science degree from Columbia College and a master's degree in sustainability and environmental management from Harvard University. He is also a graduate of the Executive Program in Sustainability Leadership at Harvard T.H. Chan School of Public Health – The Center for Health and the Global Environment and a member of the Harvard Advisory Council for the Sustainability Curriculum. He participates in many symposiums such as the Harvard Global Health Institute and University of Miami "Climate and Health," and speaks at the United Nations Conference of the Parties (COP) "Momentum for Change: ICT Solutions" in addition to presenting his previous book *Climate Health Risks in Megacities: Sustainable Management and Strategic Planning* (2016) published by Taylor & Francis Group.

List of Contributors

John P. Kotter is internationally known and widely regarded as the foremost speaker on the topics of Leadership and Change. His is the premier voice on how the best organizations actually achieve successful transformations. The Konosuke Matsushita Professor of Leadership, Emeritus at the Harvard Business School and a graduate of MIT and Harvard, Kotter's vast experience and knowledge on successful change and leadership have been proven time and again. Most recently Kotter was involved in the creation and co-founding of Kotter International, a leadership organization that helps Global 5000 company leaders develop the practical skills and implementation methodologies required to lead change in a complex, large-scale business environment.

Kotter has authored 18 books, 12 of them bestsellers. His works have been printed in over 150 foreign language editions and total sales exceed three million copies. His latest book, *Buy-In* (with Lorne A. Whitehead, 2010), focuses on the problems associated with getting others engaged and committed to good ideas and provides solutions for dealing with attacks on your good ideas. His books are in the top 1% of sales on Amazon.com.

John Kotter's articles in *The Harvard Business Review* over the past 20 years have sold more reprints than any of the hundreds of distinguished authors who have written for that publication during the same time period. Kotter has been on the Harvard Business School faculty since 1972. In 1980, at the age of 33, he was given tenure and a full professorship, making him one of the youngest people in the history of the university to be so honored.

Professor Kotter's article "Accelerate!" in *The Harvard Business Review* – in which he first debuted his Dual Operating System business model – won the 2012 McKinsey Award, which recognized it as the most significant article published in the magazine that year. This article has since been expanded into a full book by the same name.

Professor Kotter is the 2009 recipient of the Lifetime Achievement Award by the American Society for Training & Development (ASTD). This award was presented in recognition of his extensive body of work and the significant impact he has had on learning and performance in organizations.

Josef Leitmann is Lead Disaster Risk Management Specialist at the World Bank, heading teams on Resilient Recovery and Urban Resilience at the Global Facility for Disaster Reduction and Recovery (GFDRR). He is also GFDRR's focal point for humanitarian and fragility/conflict issues. Previously, Joe spent four years managing the US$400 million Haiti Reconstruction Fund, which supports post-earthquake recovery in partnership with the government and the international community. He also developed and managed the

US$650 million Multi Donor Fund to help rebuilding efforts after the tsunami in Indonesia.

Joe has more than 30 years of development experience with the World Bank in disaster risk management, climate change, natural resource management, urban development, forestry, and clean energy. He has worked in more than 40 countries and held long-term assignments in Turkey, Brazil, Indonesia, Haiti, and the South Pacific (the latter as a US Peace Corps volunteer). Dr. Leitmann holds a PhD in city and regional planning from UC Berkeley and a master's degree from the Harvard Kennedy School. He is the author of a textbook on urban environmental management, *Sustaining Cities*, co-author of the World Bank's *Investing in Urban Resilience*, and the author of numerous articles.

Eric Stark Maskin is an American economist who was awarded the 2007 Nobel Memorial Prize in Economic Science with Leonid Hurwicz and Roger Myerson for laying the foundations of mechanism design theory. The theory, initiated by Hurwicz, allows economists to distinguish situations in which markets work well from those in which they do not. It takes into account information about individual preferences and available production technologies, which is usually dispersed among many parties who may use their private information for selfish ends. Maskin and Myerson each developed the theory further, greatly enhancing our understanding of the properties of optimal allocation mechanisms in such situations, accounting for individuals' incentives and private information. The theory has helped economists identify efficient trading mechanisms, regulation schemes, and voting procedures, and it is of importance not only to economists but in several areas of political science.

Professor Maskin is the Adams University Professor at Harvard University. Throughout his extensive career he has made contributions to game theory, contract theory, social choice theory, political economy, and other areas of economics. He earned his AB and PhD from Harvard and was a postdoctoral fellow at Jesus College, Cambridge University. He was a faculty member at MIT from 1977 to 1984, Harvard from 1985 to 2000, and the Institute for Advanced Study from 2000 to 2011. He rejoined the Harvard faculty in 2012.

Robert Pojasek, president of Pojasek & Associates, is an internationally recognized expert on the topic of business sustainability and process improvement. He assists clients with developing and facilitating the planning and implementation of sustainability programs and sustainability management systems at both the corporate and facility level. Management systems help make sustainability programs a part of what every employee does every day. Dr. Pojasek has extensive experience with the implementation of a variety of management system standards, including quality, environment, occupational health and safety, corporate social responsibility, and sustainability. He utilizes combinations of conventional management systems (ISO 9001, ISO 14001, and OHSAS 18001), risk management (ISO 31000), social responsibility (ISO 26000 and AS 8303), sustainability (BS 8900), business excellence frameworks (e.g., Baldrige Performance Excellence), and process improvement (Lean and Six Sigma).

Dr. Pojasek has prepared business continuity plans and pollution prevention plans. He has also prepared corporate responsibility reports and applications for Dow Jones sustainability index recognition and is also experienced with implementing the US Environmental Protection Agency's National Enforcement Investigation Center (NEIC) compliance-focused management system and the Occupational Safety & Health Administration Voluntary Protection Program (OSHA VPP).

With more than 35 years of experience in the environmental, health, and safety consulting field, Dr. Pojasek has worked with a diverse range of clients in the manufacturing and service sectors and for non-government organizations (NGOs) and government agencies. During this time, he has been very active in the practice of pollution prevention and has presented numerous conference presentations and written 100 publications on pollution prevention and sustainability practices. As an adjunct professor at Harvard University, he teaches "Strategies for Sustainability Management," and serves as a thesis director for students conducting research in sustainability at the masters' degree level.

1

Information and Communication Technology and Sustainable Development: An Imperative for Change

ICTs in the Context of Development Goals

The progression of the information and communication technology (ICT) sector and sustainable development have accelerated through the innovation and urgent need of finding solutions to cope with global risks. A familiar place of knowledge and technology as a major feature of the knowledge-based economy is an important step to begin strategic action. Never before has humanity faced, on a global scale, natural challenges imminent and uncertain at the same time. Information technology can be both a vital tool and critical link for world nations to foster *economic* development, as well as addressing *social and environmental issues*. Therefore, it became imperative to utilize ICTs to combat environmental hazards and its repercussions for society's interest in the environment and sustainable development. This leads to actionable benefits of ICTs that can be placed in promoting economic growth, minimizing poverty, and addressing environmental challenges. Their social implications may present themselves as temporary or permanent changes to the atmosphere, water, and land due to human activities, in addition to social exclusion, poverty, substantial levels of inequalities, and health and demographic challenges, which can result in impacts that may not be irreversible. Literature concerning both the potential and challenges of ICTs in the climate change field began to emerge at the beginning of the 2000s (Ospina and Heeks, 2010), paving the way for new developments and implementations for practical solutions between information society and the environment.

From a global perspective, the relationship between ICTs, sustainable development, and the environment explores the use of ICTs in the context of development goals, expressly the realization of the Millennium Development Goals (MDG), which are eight goals with measurable targets and clear deadlines for improving the lives of the world's poorest people. Moreover, the MDGs aim at warranting environmental sustainability

(Millennium Development Goals, 2015). Research started to emerge at the end of the 1990s with an increasing awareness of environmental sustainability. Climate change wasn't addressed specifically at the time; instead, the aforementioned studies were addressing a growing concern to the negative and positive effects of ICTs in the field in addition to these technological benefits in monitoring the environment (Ospina and Heeks, 2010). Emissions of carbon dioxide (CO_2) and other gases' negative effects on the planet created a focus in technological advances to find concrete solutions. Within that field of research, the studies concentrated on the potential of ICTs toward CO_2 emission reduction, including a multiplicity of highly innovative applications cultivating important actions toward achieving energy efficiency and innovation in such areas as the telecommunication and transport sectors. The emerging evidence of the consequences of inaction to cope with climate change impacts opened the door to new uses of ICT applications in vulnerable contexts to climate change. In particular, developing countries are in a susceptible position to environmental challenges. The role of ICTs can be essential in tailoring strategic planning to mitigation and adaptation, as well as emerging applications that could help improve the access to climatic information by providing fundamental support for decision-making processes at local and national levels.

Broadband Networks to Minimize Detrimental Effects and Promote Economic Growth

There is an interlink between the gross domestic product (GDP) and the expansion of broadband networks, services, and applications where broadband penetration contributes to GDP growth and economic prosperity, although not all GDP growth equals economic equality among populations. There are several studies that corroborate the aforementioned statement, such as a 10% increase in broadband penetration adding to an extra 2.5% GDP growth in China (Broadband Commission, 2012). According to the World Bank, in high-income countries a 10% rise in broadband penetration adds a 1.21% rise in economic growth, or 1.38% for low- and middle-income countries (Minges, 2015). The Broadband Commission study (2012) estimated that the ICT sector contributes 2–2.5% of global greenhouse gas (GHG) emissions, including radio communication systems and equipment, but its largest contribution is in enabling energy efficiency in other sectors. Conversely, the overall productivity growth among OECD countries is approximately 40% of ICT's use and its applications can have the potential to reduce energy consumption in the remaining 97–98% of carbon emissions. Moreover, ICT has the capacity to implement carbon savings five times greater than the sector's own total emissions (2% of total greenhouse gas emissions) (SMART, 2020). The new report updates corroborated the ICT's potential to reduce global emissions even further with more than 7.8 gigatons by 2020—equivalent to a 15% reduction of global emissions, for only a low increase in ICT's own emissions, particularly as new services and applications enabled by high-speed high-capacity broadband are being developed (ICT's and the Internet's impact on job creation and economic growth, 2012).

There are three principal roles for the ICT sector to achieve sustainable development goals:

- *Transformation*: The transformation of *physical products* and systems to digital services reducing through dematerialization replacing travel with collaborative tools such as video-conferencing or substituting the need to produce physical products by delivering e-products and services (Broadband Commission, 2012).

- *Climate mitigation*: Greening ICT is referred to as the reduction of the sector's own emissions. Therefore, developing energy lean products with a strategic plan for reduction targets is primordial to establish concrete solutions to mitigation. Process optimization also plays an important role like smart grids, smart logistics, smart buildings, and smart motor systems (Roeth and Wockek, 2011).

- *Climate adaptation*: Aimed at the vulnerability of systems to the effects of severe weather events. Adaptation initiatives can be accomplished by weather information and disaster alerts and conducting routine risk assessments to identify and plan for high-risk situations. Based on identified risks, continuity and preparedness plans are developed and tested.

Health: The Face of Environmental Risks and Sustainable Development

ICT and the Health Sector

There is a link between changes in extreme weather and climate events' impacts on public health and the exacerbation of those impacts on the poor populations. We find implications of adaptation and mitigation strategies, and what the ICT sector can contribute to minimizing such detrimental effects on the vulnerable inhabitants. The discussions about the potential of ICTs to improve the health and well-being of poor and marginalized populations have been in the international arena of major players such as the United Nations and the World Bank. However, the aforementioned potential of ICTs has not been used effectively. Much more research is needed and, even more importantly, the effectiveness of implementing models to increase information flows and the dissemination of evidence-based knowledge to empower citizens needs to be re-evaluated and developed for better results.

The tools and approaches used in ICT have not been extensively used as instruments to advance equitable health-care access, particularly in developing countries. The applicability, relevance, and cost-effectiveness of the approaches are a relatively new stage of development and implementation. Therefore, world leaders, policy-makers, and decision-makers are hesitant to determine their investment priorities (Chandrasekhar and Ghosh, 2001). Nevertheless, it is worth mentioning some important initiatives that have proven positive results, such as a 50% reduction in mortality or 25–50% increase in productivity within the health-care system (Greenberg, 2005). Moreover, ICTs have been important in the improvement of the health sector. It has enabled:

- The communication and remote consultation;
- Dissemination of public health information and the facilitation of public discourse and dialogue around major public health threats;
- Diagnosis and treatment through telemedicine;
- Facilitated collaboration and cooperation among health workers, including the sharing of learning and training approaches;
- Supported more effective health research and the dissemination of and access to research findings;
- Strengthened the ability to monitor the incidence of public health threats and to respond in a more timely and effective manner;
- Improved the efficiency of administrative systems in health-care facilities (Ramesh et al., 2014).

There is a promising role of ehealth as a mitigation and adaptation strategy to reduce GHGs and societal vulnerability to changes in the climate system. However, a gap exists between implicit and empirically demonstrated benefits of ehealth simply because only a few studies have been conducted with limitations in opportunities for implementation. Undoubtedly, ehealth is modeled to increase quality, efficiency, and access to health care, particularly in remote areas. Initial results from the largest home health-care clinical study, the UK Whole System Demonstrator (WSD) program, strongly support these statements. Preliminary outcomes show 15% reduction in emergency room visits, 20% reduction in emergency admissions, 14% reduction in elective admissions, 14% reduction in bed days, and 45% reduction in mortality for patients with chronic heart failure (CHF), chronic obstructive pulmonary disease (COPD), and diabetes. Telemedicine is thus considered one of the few reasonable solutions to address the growing number of elderly and chronically ill people in the developed world (Holmner et al., 2012). Environmental and climate-related effects can overwhelm the health-care system and health-care facilities will need to assess the associated risks and adopt new technological tools and risk management frameworks to deter and/or adapt to the harmful impacts. The delivery of health care entails efficiency and timely delivery of services and, as health experts warn that despite years of effort toward improving emergency preparedness when disasters occur, the health system is still vulnerable to disruptions. Nonetheless, the growth of powerful new health information technologies (HITs) can enhance the delivery of health care and the promotion of health with the support of mobile health communication devices, advanced telehealth applications, access to relevant health information, enhance the quality of care, reduce health-care errors, and increase collaboration. Nevertheless, the opportunities for HIT to develop new ways of providing efficient technological tools to cope with climate health risks need to be tailored to the efficient design of mHealth applications to meet the health literacy levels of different audiences that communicate effectively with a diverse array of health-care consumers, providers, and policy-makers (Kreps, 2017).

Between 1994 and 2013, 6,873 natural disasters were recorded worldwide, which claimed 1.35 million lives or almost 68,000 lives on average each year. In addition, 218 million people were affected by natural disasters on average per annum during this 20-year period (Centre for Research on the Epidemiology of Disasters, 2015). Nonetheless, the science of weather forecasting and climate monitoring, which is critical to reducing such high casualty rates, is being advanced by the development in ICTs.

An Example of Weather Forecasting and Telecommunication Networks

Extensive weather station networks are needed for monitoring key climate parameters such as wind speed, precipitation, barometric pressure, soil moisture, wind direction, air temperature, and relative humidity. These parameters may be used both for forecasting and for decadal climate modeling. The technologies needed include weather satellites and both local and remote automated weather stations. Just as with telecommunications networks in general, there are logistical and financial problems in achieving sufficient global coverage to collect the required data. Satellite observations include visible spectrum cameras to detect storms and deforestation, infrared cameras to detect cloud and surface temperatures and sea level rise, and particle detectors of solar emissions.

The Geostationary Operational Environmental Satellites (GOES-11&12) and others are capable of making these observations, which are essential in providing input to weather forecasting and climate models. Emphasis is now on improving coverage of space and land-based sensors. Fine resolutions are needed, with frequent updates, to provide the most accurate forecasts. For example, the European Meteosat-8 located over the Atlantic Ocean at 0° longitude provides an operational European "rapid scan" mode service, which commenced in the second quarter of 2008 (with images of Europe every 5 minutes). Meteosat-9, also at 0° longitude, provides the main full Earth imagery service over Europe and Africa (with images every 15 minutes). More work is needed to establish whether Africa and other developing regions could benefit from dedicated weather satellites, with improved resolution over their regions, to match the standards of weather and climate variability and change forecasting in developed regions.

Integration and assimilation of ehealth into the everyday life of health-care workers is becoming a reality in developing as well as developed countries (World Health Organization, 2008). ICTs enable online communication about medical issues and diagnosis of complicated diseases by linking medical practitioners who are separated geographically. They have the potential to change the delivery of health-care services and patient care, as well as the management of health-care systems, which is crucial during severe weather events or situations of catastrophic disruptions. Moreover, the application of "sustainable" or "equitable" health care is another facet of ehealth, providing care that does not vary in quality because of personal characteristics such as gender, ethnicity, geographic location or socio-economic status. In the case of ethnicity, there are solutions which can assist with language translation or the use of graphics to aid communications. Geographical challenges can be overcome by emerging ehealth innovations such as telemedicine for remote consultations, teleradiology for remote expert readings, monitoring home patients and chronic diseases. Many hospitals and care facilities are benefitting from more efficient care processes, reducing waits and sometimes harmful delays for both those who receive care and those who give care. When Hurricane Sandy tore through the northeast in the last days of October 2012, it affected almost everyone in its path. In New York City, where several hospitals were evacuated due to power outages and flooding, New York's statewide health information exchange ran without a flicker, fulfilling the promise of health information technology brought into focus by Hurricane Katrina, which devastated Gulf Coast health-care organizations in 2005. Information technology applied to health and health care supports health information management across computerized systems and the secure exchange of health information between residents, providers, and quality monitors. Hence, better delivery of interventions and the use of new technologies and processes improve the way health interventions are delivered to increase climate resilience and identify health risks.

Health-Care Computing

The origin of health-care computing can be traced back in the early 1950s when it was only available in the major hospitals of G7 countries (Canada, France, Germany, Italy, Japan, the United Kingdom, and the United States) because they were able to afford the use of workstations and machines. It required a considerable effort among health professionals and computer experts to define efficient and viable uses of machines in the sector. The demand for expertise was increasing, but the end results were mostly troubled with mechanical and programming errors. From the early 1960s through the 1970s, a new era of computing in health care emerged. Nonetheless, many of the early projects were almost complete failures, as the complexity of the information requirements of a patient management system was gravely underestimated. From the early 1960s through the 1970s, a new era of computing in health care materialized and the evolution of technology in the health-care systems has been rapidly advancing as much as the challenges the sector faces, in addition to public health risks caused by climate variability and change that leads to the urgency of producing tangible results in the short and long run (Institute for Sociology, Center for Social Sciences, 2013). The increasing processing power of computers resulted in a dramatic move away from massive health data processing using mainframe or minicomputers to new and more efficient forms of health management information systems (HMIS), office automation (OA), and networking technologies (Tan, 2005). Clinical integration, health networking, and telecommunications provided benefits of IT convergence in health care while boosting patient care and service levels (OECD, 2009). The convergence of knowledge, technology, and society is a fundamental factor for sustainable development and a healthy and prosperous world. This translates into a transformative power of junction between new technologies and environmental risks by providing the fact-based wisdom to avoid detrimental consequences and the core opportunity for progress in the twenty-first century.

Improving Health and Connecting People

In developing countries, ICTs have proven to be a tremendous accelerator of economic and social progress. However, emerging trends in technologies that are likely to shape ICT use in the health sector, to facilitate communication and the processing and transmission of information by electronic means, are still in the developing stage as compared to the developed nations specifically in North America and Europe. They have a major shortage of resources, both financial as well as trained human resources for adapting such systems. Moreover, developing countries are also suffering from political, social, cultural, and other types of constraints that limit their efforts to significantly implement ICTs into the health system (Burney et al., 2010).

ICT eHealth Policy Implementation

Countries do not have specific ICT strategies to cope with risks. The ICT initiatives are disseminated throughout systems in a way that prioritizes events according to the urgency of the situation they faced and are often captured as a horizontal issue, such as improvement of the business environment, and extroversion and expansion of the economies. The urgent need of tackling harmful impacts of climatic risks and adapt to the new present and future climatic conditions, among other issues, called for strategies leading to research on *those issues* and health using ehealth as a tool. In that regard, national policies pave a way to promote the use of ehealth creating effective incentives in the health sector to develop adaptation and mitigation strategies. Therefore, the understanding of the benefits of technology by leaders in the decision-making process becomes crucial to implement green ICT initiatives to reduce carbon dioxide (CO_2) emissions accompanied by monitoring and quantifying progress in estimating emission reductions in the health sector in a way that makes international comparisons possible. The health sector specifically needs to be fully involved in the decision-making process to accelerate the transformation of ehealth as a mitigation and adaptation strategy. Using ehealth as a climate policy can be a motivator to entrust to ehealth, since it provides co-benefits such as accessibility, connectivity, and cost-savings, and improves the quality of services (Holmner et al., 2012).

List of Terms

Broadband network: A high-speed communications system that links computers to the Internet. It uses a cable, DSL, or satellite modem hooked up to an Internet service provider (ISP).

Carbon dioxide (CO_2): A colorless, odorless, and non-poisonous gas formed by combustion of carbon and in the respiration of living organisms and is considered a greenhouse gas. Emissions mean the release of greenhouse gases and/or their precursors into the atmosphere over a specified area and period of time.

Climate change: Refers to significant changes in global temperature, precipitation, wind patterns, and other measures of climate that occur over several decades or longer.

Climate change adaptation: The United Nations Framework Convention on Climate Change (UNFCCC) defines it as actions taken to help communities and ecosystems cope with changing climate condition. The Intergovernmental Panel on Climate Change (IPCC) describes it as an adjustment in natural or human systems in response to actual or expected climatic stimuli or their effects, which moderates harm or exploits beneficial opportunities.

Climate change mitigation: It refers to efforts to reduce or prevent the emission of greenhouse gases. Mitigation can mean using new technologies and renewable energies, making older equipment more energy efficient, or changing management practices or consumer behavior.

Cyberterrorism: According to the U.S. Federal Bureau of Investigation, cyberterrorism is any "premeditated, politically motivated attack against information, computer systems, computer programs, and data which results in violence against non-combatant targets by sub-national groups or clandestine agents" (Khanka, 2014).

Distributed Denial of Service (DDoS): An attempt to make an online service unavailable by overwhelming it with traffic from multiple sources.

ehealth: An emerging field in the intersection of medical informatics, public health, and business, referring to health services and information delivered or enhanced through the Internet and related technologies. In a broader sense, the term characterizes not only a technical development, but also a state-of-mind, a way of thinking, an attitude, and a commitment for networked, global thinking, to improve health care locally, regionally, and worldwide by using information and communication technology.

Greenhouse gas (GHG) emissions: A gas in an atmosphere that absorbs and emits radiant energy within the thermal infrared range. This process is the fundamental cause of the greenhouse effect. The primary greenhouse gases in Earth's atmosphere are water vapor, carbon dioxide, methane, nitrous oxide, and ozone.

Known unknowns: Result from phenomena which are recognized, but poorly understood.

Health information technology (HIT): Information technology applied to health and health care. It supports health information management across computerized systems and the secure exchange of health information between consumers, providers, payers, and quality monitors.

Information and communication technology (ICT): Refers to technologies that provide access to information through telecommunications. It is similar to information technology (IT), but focuses primarily on communication. This includes the Internet, wireless networks, cell phones, and other communication mediums. For the purposes of this primer, ICT is a diverse set of technological tools and resources used to communicate, create, disseminate, store, and manage information.

Millennium Development Goals (MDGs): Eight goals with measurable targets and clear deadlines for improving the lives of the world's poorest people. To meet these goals and eradicate poverty, leaders of 189 countries signed the historic Millennium Declaration at the United Nations Millennium Summit in 2000.

Sustainable development: Development that meets the needs of the present without compromising the ability of future generations to meet their own needs.

Sustainable Development Goals (SDGs): The SDGs are a collection of 17 global goals set by the United Nations. The broad goals are interrelated though each has its own set of targets to achieve. The SDGs are also known as "Transforming our World: the 2030 Agenda for Sustainable Development" or 2030 Agenda in short.

Telehealth: A collection of means or methods for enhancing health care, public health, and health education delivery and support using tele-communication technologies.

Unknown unknowns: (unexpected or unforeseeable conditions), which pose a potentially greater risk simply because they cannot be anticipated based on past experience or investigation.

2

Using Information and Communication Technologies (ICTs) to Tackle Climate Variability and Change Impacts

The Role of Information Communication Technologies in Combating Climate Variability and Change

The advance of information and communication technologies (ICTs) implementation in finding solutions to our socio-economic and environmental issues is playing an important role in achieving sustainable development goals. The link between ICTs and policy-makers and all stakeholders of the information society is more important now than ever, as the capacity of new technologies to cope with climate variability and change as well as addressing health risks; environmental degradation becomes crucial to minimize the aforementioned global risks. Hence, the development of pragmatic policies to mitigate the harmful effects of climatic events imposes a responsibility to promote the technology as an effective tool for developing and consequently implementing strategic planning to massive ICT infrastructure investments and availability of broadband for local and national development. To further the efforts of innovation and coordination requires the participation of all sectors of society; from the public and private sector to communities and local leaders in order to fulfill the potential that ICTs offer toward a sustainable society.

Climate variability and change carry a wide range of stressors that poses potential detrimental effects to populations including heavy rainstorms, cyclones, heat waves, sea level rise, extended periods of flooding or drought, changing patterns of temperature and rainfall, among others. The assessment of vulnerabilities within the context of population density, infrastructure, network, transportation, and socio-economic factors that can alter the stability of the region affected is highly important and needs to be analyzed to understand the threats and opportunities to cope with the aforementioned risks. We need to highlight that not all risks are harmful. Some risks present an opportunity to better existing situations and even improve them.

The use of ICTs in developing countries have been explored in recent years as the impacts of climate variability and change are more pronounced and the poorest countries lack adequate infrastructure, health-care systems, and telecommunication networks, which makes them more vulnerable and requires urgent action toward sustainable solutions.

Information and Communication Technologies as Enablers of "Networked Governance"

Process governance is a major issue and must be integrated into the adaptation and mitigation strategies. It encourages the alignment of initiatives in a way in which an organization can consolidate the *process* management initiatives within standards, rules, and guidelines. Governance processes to foster innovation play a key role in mitigation and adaptation strategies with the inclusion of multi-stakeholders and the understanding of climate-related concerns. The processes are to obtain sustainable development goals and to develop oriented concepts to build suitable systematic review and frameworks related to their use in order to monitor, measure, and assess climate change. The indirect effects are those emerging from the use of ICTs to increase awareness and facilitate public dialogue (e.g., via Web 2.0 and social networking), systemic effects using ICTs as enablers of "networked governance," keys to adapting to climate change, and achieving sustainable development (MacLean, 2008).

These connections are illustrated through the examples that follow and in Table 2.1.

TABLE 2.1

Food Security Information and Communication Technology Potential Examples and Expected Impacts

Sub-Sector	Examples of ICT Potential	Expected Impact on Food Security
Agriculture	Radio programs can be an effective tool in remote rural areas for the dissemination of knowledge and information on improved land management practices (e.g., improvement of soil fertility and structure)	ICTs can help to raise awareness and create new capacities on improved land management practices, which can translate into production benefits (e.g., higher crop yields)
	Mobile phone text messages (SMS) can be sent to farmers in support of integrated nutrient management programs (e.g., sending SMS reminders on when to apply fertilizers)	ICTs can facilitate continuous monitoring and support from experts in the implementation of agricultural practices, including precision farming.
	Internet-based applications (e.g. remote sensing, GIS, climate change models, data mapping) can be used in support of agricultural planning, helping farmers to allocate resources more effectively and reduce risks	ICTs can help to reduce uncertainties generated by climate change through relevant information that, if presented in appropriate formats and adequate scales, can inform farmers' decision-making

(Continued)

TABLE 2.1 (CONTINUED)

Food Security Information and Communication Technology Potential Examples and Expected Impacts

Sub-Sector	Examples of ICT Potential	Expected Impact on Food Security
	Participatory videos can allow communities to document their experiences using traditional and new seed varieties under changing climatic conditions, to share lessons learned and to foster appropriate crop selection (e.g., drought/flood or saline tolerant)	ICTs can foster crop diversification by helping to document and share traditional knowledge and experiences with resilient seed varieties
Livestock	Videoconferences with experts, held in community access centers (e.g., Telecenters) can facilitate the access to information, knowledge and technical advice without having to travel to other villages or towns. This includes video and e-mail-based consultations on improved feeding and nutrition practices, animal health control and grassland management practices under changing climatic conditions	ICTs can facilitate access to expert technical advice to complement local knowledge and point livestock owners to alternative practices, contributing to animal productivity under situations of climate stress (e.g., providing advice on genetics and reproduction, grazing schedules or supplements for poor quality forages)
Fishery	Internet and community radio can be used to create awareness and provide access to content on fisheries codes (e.g., code of conduct for responsible fisheries) and regulations, as well as information on aquaculture management in different climatic conditions (e.g., feeding practices, selection of stock)	ICTs can enable access to user-friendly (e.g., using local languages, images, and sound) regulatory content (e.g., policies, rights and obligations) that can help inform decision-making and management approaches, having an impact on fish productivity and sustainability
Agroforestry	Mobile technologies(e.g., smartphones and PDAs), Web 2.0 and social media applications (e.g., Facebook and Twitter) can be used to collect and disseminate information on the use of trees and shrubs in agricultural farming systems (e.g., sharing advantages of growing multipurpose trees, alternatives of plantation/crop combinations, the use of live fences and fodder banks in contexts affected by climatic variability)	ICTs can help to motivate stakeholders toward the adoption of agroforestry practices to increase farm incomes and diversify production. ICTs can also help to gather and mobilize stakeholders for local conservation actions

Source: Gupta (2013).

Enhanced food security and resilience became a primordial issue that is addressed to a series of factors described below:

1. Place importance on the role of players or "infomediaries" that gather and link information such as extension workers, locally trained professionals, and youth at the local level, supporting the use of ICTs for climate change awareness, mitigation, monitoring, and adaptation. They can be the facilitators to build the bridge between the scientific knowledge and technical climate change data bringing pragmatic applications (Ospina and Heeks, 2012).

2. Build the capacity of local stakeholders and embrace the ICT solutions' potential to exchanging technical information, interlinking agricultural networks and experts who exchange information to improve processes with local and external players, as well as identifying relevant information to minimize risks.

3. Emphasize the role of policy-makers and create awareness on the importance of ICT tools integrated into strategic approaches that aim at addressing the multiple stressors threatening food and nutrition security at the local level and understanding of the interlinkages in the water, energy, and food nexus (Artioli et al., 2017).

4. Despite the high levels of connectivity in developed countries and the growing access in developing countries, digital inclusion in rural areas remains a strong concern for policy-makers (Correa et al., 2016). Remote rural areas have many challenges in connectivity and accessibility. These challenges are noticeable in developing countries' farmers, fishers, herders, and foresters that needs to have access to ICTs services in order to take important steps to adapt and mitigate the strategies' implementation.

For rural areas in developing countries, the dissemination of information with the help of ICT tools is not sufficient to create the right environment for a strategic approach toward climatic risks. The challenge is the availability of knowledge and information reaching the appropriate stakeholders, local audiences, and agricultural producers that are able to take proper action in order to strengthen their livelihoods. In this case, ICT tools and its approach to a strategic action contribute to inform the decision-making processes of local actors and create an environment of collaboration that is essential to more resilient, food-secured agricultural systems.

Helping Communities in Developing Countries Facing Water Stress Adapt to Climate Change

ICTs have proven useful in tailoring responses to situations arising out of the climate-water nexus. ICTs have emerged as a strong way to understand water security challenges (Finlay and Adera, 2012). They are increasingly being adopted as key decision-support mechanisms for adapting to climate change effects in the developing world. However, ICTs must not be considered simply the only solution to cope with water security. Other tools must be considered that can be efficiently used in developing appropriate responses to problems addressing water issues in the climate change arena.

Sustainable Development and the Spread of Information and Communication Technology and Global Interconnectedness

The Human Progress Dilemma

The goal of accelerating human progress and achieving sustainable development goals (SDGs) is aligned with curbing the digital divide and developing knowledge societies and promoting scientific and technological innovations across areas as diverse as medicine, economy, health, and energy. Therefore, ICTs are essential to accomplish the 17 Sustainable Development Goals and empower economic growth, access to education resources and health care, and services such as mobile banking, e-government and social media, among others (Leaving no one behind: The Sustainable Development Goals Report, 2016).

The three main aspects of ICT-based empowering solutions for SDGs achievement:

1. Access to advanced technologies has grown at a fast pace and yet the impressive gains are still hampered by existing gaps in ICT access – between and within countries, between urban and rural settings, among men and women, and boys and girls. A major digital divide is still in place, with more people offline than online and particularly poor access in vulnerable developing nations. The challenge now is to bring the rest of the world online and ensuring that no one is left behind.

2. The Global e-Sustainability Initiative (GeSI) report "System Transformation – How digital solutions will drive progress towards the Sustainable Development Goals," focuses on the key findings of the study and the GeSI commitment to make the SDGs the central framework of its activities up to 2030, including stakeholder partnerships. The project includes analyses of not only the opportunities linked to ICT for the SDGs, but also the measures needed to overcome the current barriers to uptake (How Digital Solutions Will Drive Progress towards the Sustainable Development Goals, 2016).

3. The mobile industry has a critical role in promoting, advancing, and measuring the SDGs. From collaborating on accelerating a data

revolution for sustainable development to ensuring that tools and applications are developed with vulnerable communities in mind so that no one is left behind, the mobile industry's commitment to work together will be a game changer in leveraging the socio-economic impact of mobile technologies on individuals, businesses, and governments around the world (ICTs as a catalyst for sustainable development, 2016).

The adoption of ICTs is still a challenge, particularly in developing countries with billions of currently unconnected individuals. Engagement of main actors and multi-stakeholder cooperation to break those barriers and to deliver a range of life-enhancing and life-changing services at the required speed, scale, and complexity of transformation is crucial to attaining the SDGs by 2030 (ICTs as a Catalyst for Sustainable Development, 2016).

Accelerating Human Progress: The Digital Divide and Knowledge Societies

The gap between demographics and regions that have access to modern ICT, and those that don't or have restricted access is a serious concern to achieving sustainable development. The digital divide is often conceptualized as inequalities of access to technology (Thatcher and Ndabeni, 2011). The main discussion about the digital divide is emphasized on the question of the availability of technology. The application of the digital transformation is tailored to the possibility that technologies can transform political and economic systems, as well as the human agency. Many opportunities and challenges for the information society rest in economic theories of growth that try to describe knowledge-based economies, becoming pivotal for creation of resilient economies and convey increased democracy, as democracy will also be transformed in the process. There are struggles in the implementation of theories and the adoption of ICTs tools; most knowledge-related investments and execution of pragmatic solutions to achieve sustainable development are slow, particularly in several developing countries – notably in Africa, but also Latin America and South-East Asia – that continue to show low values of connectivity with low level of Internet usage and limited development of e-commerce (Bilbao-Osorio, 2013).

The digital divide exacerbates existing inequalities between the more vulnerable societies, particularly in the developing world, that leads to an uneven economic playing field, lack of financial opportunities, exposure to environmental risks, and societal isolation from restrictions to online access due to poverty-low income, cultural and political beliefs. Nonetheless, more attention needs to be made to places in rural areas, which are more inaccessible to the digital transformation that affect communities and regions; where the main argument for closing the digital divide is more prominent and requires a stronger urgency for execution by bridging the gap and creating a "knowledgeable-interconnected society" for sustainable growth and prosperity. In 1985, a special commission of the International Telecommunication Union (ITU) released what is commonly known as "The Maitland Report," formulating the positive impact of telecommunications as "an engine of growth and a major source of employment and prosperity." The report highlighted the ICTs benefits, primarily in developed economies (Pepper and Garrity, 2015).

The potential of ICTs as an enabler for socio-economic prosperity is greater than ever and the latest available data from the World Bank show income inequality (the distribution of income across all people in the world) to be on the decline. ICTs can be used to directly influence the productivity, cost-effectiveness, and competitiveness in industries. Developing countries can maximize these opportunities and build a robust economy utilizing ICT tools, allowing opportunities and minimizing the threats which exist in the context of globalization. The global income inequality has fallen steadily from a Gini coefficient of 72.2 in 1988 to 70.5 in 2008 (Pepper, 2015). They attribute the decrease in inequality to the large overall income gains around the global median (50th percentile) of the population. The global top 1st percentile also realized significant income gains over this period, but the former (gains around the median) outweigh the latter (Pepper and Garrity, 2015).

The Impact of Internet Technologies on Poverty Alleviation

Poverty is more than just material deprivation. It also involves aspects such as lack of access to quality schooling and health care, vulnerability in the face of external events, or being excluded from decision-making processes. The contribution of health and education to poverty eradication was reconfirmed by the Human Development Report 2010 of the United Nations Development Programme (UNDP). The report states, "Countries became top performers on the Human Development Index through two broad routes, but more often through exceptional progress in health and education than through growth" (Human Development Report, 2010).

Advocating Internet diffusion and connectivity can deliver a positive impact on labor markets regardless of adoption; therefore, new technologies are having positive externalities, presenting benefits for adopters and non-adopters. The concern of the replacement of skilled labor for standardizing production processes, as weaving and spinning machines did in the early days of the industrial revolution (Goldin and Katz, 1998), is valid to a certain extent. Other technologies increase the demand for educated workers, thus placing a premium on skilled labor (Human Development Report, 2010). Despite increasing Internet access and mobile services, less than 10% of all communication carried out is related to business or developmental issues. Developing countries and rural areas in particular are suffering the most in lacking access to the Internet. Hence, people's accessibility to use the Internet is critical for sustainable human and economic development because it cannot be measured based only on the accelerated technological advances. The main reason for ICTs to support and endure sustainable development goals is to foster new forms of innovative development that are more inclusive, participative, and just (Figuères et al., 2014). The benefits and all-inclusive value of ICTs as a powerful tool for poverty eradication are not yet fully understood and developed. The question is: "How to have an objective approach to poverty (or the welfare approach) that aims at assessing what poverty is and how to get people out of it using ICTs as a catalyst tool for change?" The health sector can contribute with some answers in that regard. Social innovation in the health sector could potentially result in a completely new mindset regarding traditional methodologies to cure people of illnesses by utilizing ICTs to gather better information and to disseminate data more efficiently.

Rural hospitals, often situated in distant areas, will stand a better chance of employing highly skilled medical workers as the workers have more possibilities to connect and exchange information with their peers, feel supported and continuing to learn from consulting other professionals in their respective field (Figuères, 2013) (Table 2.2).

There are statistical analyses that combine the results of multiple scientific studies within the economic literature that demonstrate the causal impact of ICTs on economic growth. Four main mechanisms dictate the process by which ICTs contribute to macroeconomic growth by affecting inputs to the gross domestic product (GDP) growth:

1. ICTs contribute to GDP directly through the production of ICT goods and services as well as through continuous advances in ICT-producing sectors.
2. ICTs contribute to total factor productivity growth through the reorganization of the ways in which goods and services are created and distributed.
3. ICT industries generate positive employment effects.
4. Increasing applications of ICTs (capital deepening) leads to rising labor productivity.

With ICTs contributing to global economic growth, developing regions have experienced a steady decline in absolute poverty. The global extreme poverty rate (those individuals surviving on less than $1.25/day) has dropped from 1.9 billion people in 1981 to 1.3 billion in 2010 according to the World Bank; a drop in extreme poverty rates from greater than 50% to 21%. The decline in extreme poverty is led by developing countries' growth in some African nations, China, and India. Moreover, social programs in Latin America contributed to the economic progress as well despite the noticeable inequalities across different sectors of society (e.g., low income, gender, etc.).

TABLE 2.2

The Growth Effect of Internet Technologies

Authors	Data	Methodology	Key Findings
Gillett et al. (2006)	Panel of US zip codes and counties for 1998–2002.	Instrumental variable regression and matched sample.	Availability of broadband adds between 1% and 1.4% to growth rate of employment and 0.5–1.2% to growth rate of business establishments.
Crandall et al. (2007)	Panel of US stales for 2003–2005.	OLS regression.	A 1% increase in broadband penetration yields an increase of between 0.2% and 0.3% in employment rate. No effects on GDP growth.
Qiang and Rossotto (2009)	Panel data of 120 countries for 1980–2006.	OLS regression.	For high-income economies: a 10 percentage point (p.p.) increase in broadband penetration yields a 1.21 p.p. of additional GDP growth. For developing countries: a 10 p.p. increase in broadband penetration yields a 1.38 p.p. of additional GDP growth.
Koutroumpis (2009)	Panel of 22 OECD countries for 2002–2007.	Simultaneous equations model with instrumental variable.	A 10% increase in broadband penetration rate increases economic growth by an average of 0.25%.
Czernich et al. (2011)	Panel of 25 OECD countries for 1996–2007.	Instrumental variable regression.	A 10 p.p. increase in broadband penetration raises annual GDP per capita growth by 0.9–1.5 p.p.
Mayo and Wallsten (2011)	Panel of US stales for 2006–2008.	OLS regression.	Small negative effect of broadband penetration on GDP growth and employment.

Source: Galperin and Viecens (2017).

Social, Virtual, and Physical Mobility: ICT will define social mobility. The social impact of ICT is evident as it influences physical travel, real-time situations, the rapid spread of information, 'tele-offices' and tele-commutes, online platforms, and networking sites promoting earlier, richer and more meaningful engagement with the world. Virtual mobility is heavily present with new advancements in technology within the educational system. The role of ICT in providing the tools for developing communication, cognitive, labor, and business skills and competencies to help gather information (knowledge) is growing. ICTs also help acquire skills and abilities creatively through online platforms; further, applying them in practice for achieving teaching and learning goals (Rutkauskiene and Gudoniene, 2015).

ICT is a moving target: Empirical transport researches hardly keep up with present-day technological and commercial developments. Nevertheless, the availability of smartphones – and with them, the mobile Internet usage – is a global phenomenon that provides Internet access to billions of users. One of the most basic elements of human behavior and social organization is to gather, process, and disseminate information and ICTs enhance the capabilities of people to carry these powerful drivers of behavioral and social change (Pawlak et al., 2015).

Democracy: The use of the Internet and access to information can lead to a healthier democracy; it can spread communication about social and economic unrest, educate and inform the population more accurately about actual events, increase public participation in elections and decision-making processes. Therefore, ICTs pose opportunities and also threats to democracy. The promise of an information-rich society, in which citizens have access to a wide range of information from a variety of sources; where debates among citizens and policy-makers with full participation in the political process is greatly increased. However, ICTs also threaten to undermine democracy by compounding existing biases in the distribution of knowledge and information and reducing participation among the distanced and marginalized votes. New ICTs, therefore, have ambiguous but profound consequences for democracy, both now and in the future (Horrocks and Pratchett, 2009).

It is noteworthy to address the fact that countries with high-income inequality translates into low social mobility. ICT diffusion can affect progress that allows every member of a society to live in an environment with high economic, political, and civil liberties and those outcomes differ accordingly to the income level, accessibility, and type of technology. It is noteworthy to mention that measuring human development entails the implications of environmental and socio-economic factors that improve or alter the sustainable development goals that are tailored to effective policies and strategies. The components of human development, which encompass the process of enlarging people's freedoms and opportunities and improving their well-being, are presented in the columns in Table 2.3, (health, education,

TABLE 2.3

Measuring Human Development

| Empirical Measure | Components of Human Development | | | | |
	Health	Education	Material Goods	Political	Social
Average level	Human Development Index			Empowerment indicators	
Deprivation	Multidimensional poverty index				
Vulnerability	Indicators of environmental sustainably, human security, well-being, decent work				
Inequality	Inequality-adjusted HDI				
	Gender Inequality index				

Source: Human Development Report (2010).

material goods, political participation and social cohesion), and the rows list the empirical measures of those components (deprivation, average level, vulnerability, and inequality).

Human development is a broad concept that encompasses all the fields of human social life and sustainable development is what human development advocates. Economic growth is presented as the panacea that can solve any of the world's problems, but it must be aligned to sustainable ecosystems assuring a balance between human development, achieving the well-being of the world population, and ethical universalism. This is basically "an elementary demand for impartiality – applied within generations and between them" providing the basic elements of a sustainable and prosperous world (considering the economic, social and environmental responsibilities of society as a whole) and arguing "in favor of giving priority to the protection of the environment as the ethical need for guaranteeing that future generations would continue to enjoy similar opportunities of leading worthwhile lives that are enjoyed by generations that precede them" (Anand et al., 2000). Sustainable development strategies need to prioritize human development – focusing broadly on improving the quality of human lives rather than narrowing on the richness of economies. Any economic model has to be tailored to a sustainable development model, which is a multidisciplinary concept and it relies on reducing resource consumption, producing clean alternative energy, protecting environmental factors and the quality of life in its complexity (Elena Raluca Moisescu, 2015).

List of Terms

Climate variability: The way climate fluctuates yearly above or below a long-term average value. It is not as noticeable as weather variability because it happens over seasons and years.

Digital divide: A term that refers to the gap between demographics and regions that have access to modern information and communications technology (ICT), and those that don't or have restricted access. This technology can include the telephone, television, personal computers, and the Internet.

Gini coefficient: In economics, the Gini coefficient (sometimes expressed as a Gini ratio or a normalized Gini index) is a measure of statistical dispersion intended to represent the income or wealth distribution of a nation's residents, and is the most commonly used measurement of inequality.

Global e-Sustainability Initiative (GeSI): A strategic partnership of the Information and Communication Technology (ICT) sector and organizations committed to creating and promoting technologies and practices that foster economic, environmental and social sustainability.

Gross domestic product (GDP): The monetary value of all the finished goods and services produced within a country's borders in a specific time period.

Human development: The process of enlarging people's freedoms and opportunities and improving their well-being.

Networked governance: A dynamic process of organizing rather than a static entity.

Process governance: Comprises the definition of overall guidelines of the process management model, the process control model and the activities of the various organizational units, and involves mainly the distribution of process management-related responsibilities within the organization. Briefly, it involves fostering the definition of overall guidelines to orient what should be done in process management and how it should be done.

3

Information and Communication Technology and Sustainability

Information and Communication Technologies for Environmental Sustainability

Creating a Sustainable World

The 1987 report entitled "Our Common Future," the Brundtland Commission defined sustainability as "development that meets the needs of the present without compromising the ability of future generations to meet their own needs" (World Commission on Environment and Development, 1987). Information and communication technologies (ICTs) have great potential for environmental sustainability and its applications support the transition toward a circular economy by raising awareness on ICTs and advocating for innovative solutions that can ensure a *sustainable* future. The importance of ICT tools to achieve sustainable development goals (SDGs) and create a sustainable world is a real and pressing one (Hilty et al., 2006).

Countries across the globe can bridge the digital divide utilizing ICTs that can be used to cope with the boundaries between knowledge and action by stimulating and creating an environment of credibility and legitimacy of the information they produce (Cash et al., 2003).

Furthermore, at the organizational level, ICTs can lead the sustainability assessments to reduce risks, CO_2 emissions, e-waste, implement cost-savings programs, and manage sources effectively to have a positive effect on investment and society. There are also significant opportunities for improving environmental sustainability through ICTs in areas of energy management in housing and/or business facilities, making passenger and freight transport more efficient, and enabling a product-to-service shift across the economy (Erdmann et al., 2004).

Economic, Social, and Environmental Challenges for Sustainable Development

To achieve SDGs requires the activity and reinforcement of many sectors of society addressing the economic, social, and environmental challenges. Harvard University professor Robert Pojasek laid down a clear concept of the three responsibilities the public and private sector entities need to hasten in order to achieve sustainability and support a system of sustainable growth and development. Organizations must act in a way that is consistent with and supportive of the survival of the physical environment and also the communities and economies in which it operates (Pojasek, 2010). Organizations seeking to operate sustainably consider how their policies, strategies, and operations affect each of the following:

- Their financial performance and the broader economic system;
- The environment including the availability of resources and materials;
- People in the global and local communities in which they operate (Pojasek, 2010).

The targets of sustainable consumption and development require strong contributions across the economic, social, and environmental sector. Environmental informatics can be applied to the strategic approach in the creation, collection, storage, processing, modeling, interpretation, display, and dissemination of data and information. Policy-makers are missing opportunities from the environmental informatics community in implementing efficient strategies to cope with economic, social, and environmental risks and addressing new challenges such as scarcity of natural resources, biodiversity loss, climate change, air pollution, and the land-water-energy nexus. In essence, addressing the three responsibilities of companies puts a major focus on sustainable development with a long-term view of profitability and stability. Healthy communities where organizations operate are fundamental for a reciprocal benefit of services and products exchange, ensuring the availability of resources in a sustainable way that provides competitive advantages. The survival of the physical environment and also the communities and economies in which it operates is rapidly gaining a great importance. The need to continuously align policies, resource management strategies, and implementation directives to current economic, social, and environmental challenges is critical to support sustainable development and endure the prosperity of organizations and communities.

Information and Communication Technology in Energy Consumption and Efficiency

ICT applications in different sectors enable energy savings, increases energy efficiency, and reduces greenhouse gas (GHG) emissions. In four selected areas, the relevance of ICT for the reduction of GHG emissions was elaborated in more detail:

Electricity Distribution Grids (Smart Grids) – ICT infrastructure can play an important part within smart grids. One of the main targets for efficient operation is the reduction of transmission lines and improved performance through an advance control system. In many cases, the present distribution system is inadequate for information storage and data sharing by using real-time ICT. Therefore, the Distributed System Operators (DSO) are unable to estimate the exact system condition, that is voltage, current, and power flows. The implications of antiquated systems, which were designed for fossil-fueled power plants, are unreliable networks that make them less secure, scalable, and efficient. Large-scale applications of smart meters are being developed in many countries. The UK, for example, is committed to the full deployment of smart meters by 2020 (Masood et al., 2015).

Smart Buildings, Smart Homes, and Smart Metering – A "smart building" integrates major building systems onto a common network and shares information and functionality between systems in order to improve building operations. The full benefits of an integrated, enterprise-wide smart building strategy rely on improved energy efficiency, enhanced operational effectiveness, and a supported quality of living. Across Europe the largest emitters of greenhouse gas and energy consumption are buildings. European buildings are contributing with 40% of energy use and emissions, and US buildings accounting for 48% of the corresponding number (Toolkit on Environmental Sustainability for the ICT Sector, 2012).

The interest in smart grids is also reflected in smart homes and its benefits. Benefits include comfort and availability; control of home features such as lighting, which is an integral part of the building; security features; and control of the heating and cooling in the home, all through the use of both time and parameter-based functions. Power supply to all appliances in the home could be controlled using the smart system. Market penetration of the corresponding technologies to use these features among citizens is important in order to increase energy efficiency and minimize emission caused by fossil fuels. Nevertheless, the commercialization of these products and services needs to be accompanied by political settings such as emission standards or incentives for renewal energy. Production and demand will be one of the factors for mass consumption that can trigger new technologies in global markets with a realistic market penetration. It is worth mentioning that consumers' behaviors play a role in minimizing energy consumption. There is broad evidence in the literature that consumers save energy when aware of their consumption pattern. The integration of renewal energy sources and the releasing of data for distribution operators optimize grid management. Furthermore, smart meters may interact with intelligent appliances to allow for network-driven load-shifting activities. The reasons for the implementation of 'smart' metering differ in relation to circumstance and place. In some countries, the business case for establishing an advanced metering infrastructure (AMI) is related to the improvement of consumption feedback to customers and assisting in the transition to lower-impact energy systems (Darby, 2010).

Transport and Dematerialization

The economy has a fundamental element in the transportation of goods and services as it provides the necessary mobility and distribution of sources for society to efficiently function. Although the expansion of roads and infrastructure provides a viable way to make mobility of transportation a reality considering the economic impact and population growth, particularly in megacities, many challenges arise as traffic congestion, pollution, noise, accidents, increased reliance on fossil fuels, energy, and productivity loss. Transport and logistics are now responsible for up to 40% of the air pollution, and regulation is being implemented to reduce transport's impact on CO_2, NOx, and other emissions (Mulligan, 2014).

Moreover, urbanization is one of the most important issues of the twenty-first century. Associated with them are challenges that world leaders must confront: poverty reduction and sustainable development. These must remain the center of society's global priorities (Marolla, 2017). According to the International Energy Agency (IEA), urban areas now account for more than 71% of energy-related global greenhouse gases. This percentage, which is linked to the growth of urbanization and population density, will increase to 76% by 2030. Thus, energy-related emissions are the largest single source of GHGs when looking at allocated allowances for the areas in question (Hoornweg et al., 2011; Marolla, 2017).

Hence, the rising global population and environmental challenges due to urbanization come under significant pressures for nations and the "delivery" of transportation and its multiple applications needs to be transformed with the role of ICTs providing an integrated transport solution (Figure 3.1).

The word *dematerialization* is often broadly used to characterize the decline over time in weight of the materials used in industrial end products. One may also speak of dematerialization in terms of the decline in "embedded energy" in industrial products. Colombo (1988) has speculated that dematerialization is the logical outcome of an advanced economy in which material needs are substantially satiated. Williams (1987) has explored relationships between materials use and affluence in the United States. The main cores of dematerialization in the environmental management field lies on the minimization of waste generated, recycling and reuse of products within the production and consumption phases defining the circular economy, extracting the maximum value from them whilst in use, then recover and regenerate products and materials at the end of each service life (Williams et al., 1987).

The following graph is an example of how dematerialization works. In *Making the Modern World: Materials and Dematerialization* (2013), a comprehensive researched statistical profile of global material use, author Vaclav

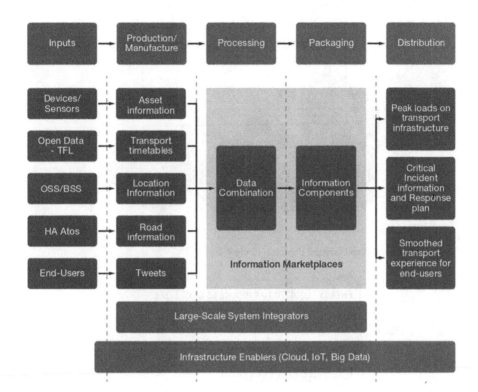

FIGURE 3.1
Information marketplace for transport in cities. (From Holler et al., 2014.)

Smil lays out just how much stuff we need to live modern lives. Material intensity continues to fall dramatically. In the United States the amount of resources extracted per dollar of GDP has decreased by nearly 75% over the past 90 years (Graph 3.1).

Energy intensity, which is the portion of the total energy supply required to produce a material, has also dropped markedly. For example, the manufacture of 1.5 gigatons of steel would have gobbled up one-fifth of the world's total primary energy supply (TPES) in 1900. In 2010 it used only about one-fifteenth (Graph 3.2).

The reductions of consumption of natural resources are significant and have the potential to maintain and increase economic growth, and contribute to the social, environmental, and consequential economic responsibilities of organizations. I must emphasize that the role of ICT development needs to be tailored to environmental policy with comprehensive critical and analytical research motivating scientific and policy debate in the field, drawing from empirical evidence. The challenges are mounting as population growth in addition to environmental degradation and increased consumption keep accelerating. Therefore, local policy planning and public and private technology funding can be an important driver for sustainable information

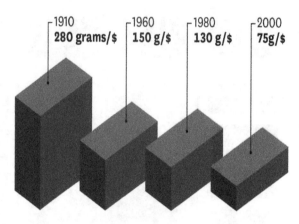

GRAPH 3.1
Energy intensity. (From Harvard Business Review, 2015.)

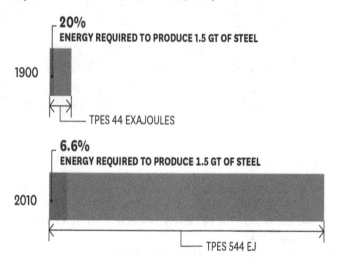

GRAPH 3.2
Energy required to produce steel, 1900–2010. (From Harvard Business Review, 2015.)

technology development. It is vital to conduct an assessment of technology adoption and determining more precise ecological sustainability objectives for ICT development (Heiskanen et al., 2001).

Information and Communication Technology and the Low-Carbon Economy

Organizations in the ICT sector are taking leading actions toward a low-carbon economy (LCE) and reducing greenhouse gas emissions. Many global organizations are taking the lead in a series of actions demonstrating that the sector is ready to put its money where its mouth is. Given its high energy demand, the ICT sector is still a net source of global greenhouse gas (GHG) emissions. The data centers used to power digital services now contribute approximately 2% of GHG emissions – on par with the aviation sector. ICT has the potential to reduce GHG emissions by 20% by 2030 through helping companies and consumers to more intelligently use and save energy. The SMARTer2030 report, an effort of the Global e-Sustainability Initiative (GeSI), which has been researching the role ICT can play in cutting global carbon dioxide equivalent (CO_2e) emissions and promoting a more sustainable society since 2008, outline three key areas of action: policy-makers, business, and consumers (SMARTer2030.gesi.org, 2015).

Enabling Environment for Information and Communication Technology

The Case for Policy-Makers

The opportunity to cut emissions globally is presented by GeSI's SMARTer2020 report. The utilization of ICT's actions such as video-conferencing and smart building management could reduce global emissions by nearly 17% compared to the 2020 forecast, amounting to US$1.9 trillion in gross energy and fuel savings and a reduction of over 9 gigatons (Gt) of CO_2e. This was the equivalent of more than seven times the ICT sector's emissions at the time. The follow-up report, SMARTer2030 (2015), is the first report of its kind to analyze the impact that ICTs have on the economy, society, and the environment.

Some of the important findings of the report within the countries and sectors examined are the savings of:

- 25 billion barrels of oil;
- 332 trillion liters of water;
- 91 million tons of paper;
- 1.6 billion people having access to ICT-enabled health-care services;
- half a billion having access to e-learning solutions;
- overall, ICT could enable a reduction of 12Gt of CO_2e from the global economy by 2030, maintaining CO_2 emissions at 2015 levels.

We need to emphasize that specific policy requirements are needed to enable these benefits. In addition, developing and developed countries, and cities that are viable of implementing such policies, should have a clear interest from authorities to exploit the methodology of assessing the impact that demand and access management strategies have on CO_2 emissions.

SMARTer2030 report highlights the need to incentivize investments in key ICT infrastructures and to promote a fair, balanced, and consistent regulatory approach to ICT. We need the policy-maker's leadership in promoting innovation and investment and to protect intellectual property rights and ensure consumer privacy and security (SMARTer2030.gesi.org, 2015).

The Case for Business

The digital transformation we are experiencing present opportunities for business growth. ICT-enabled business models are the signature of organizations' innovation and success in recent years. Many companies such as Uber and Airbnb, as well as the integration of smart solutions into people's working lives, homes, and public services, are examples of the disruptive trend in technology and society overall. These business environments also present challenges and obstacles for some, but the benefits and financial gain within a global context of ICT's impact on business and the economy are greater than ever.

The SMARTer2030 report highpoints the following actions for businesses:

1. Explore how ICT-enabled solutions can open up new sources of revenue and growth opportunities for your core business while delivering societal or environmental value.

2. Explore the potential of sector-specific cost savings made possible by ICT-enabled solutions and, likewise, be willing to collaborate and work within your sector and/or adjust your business model to realize them.

3. Commit to bolder action on emissions mitigation within your own organization and explore ways ICT can help you to meet established targets (SMARTer2030.gesi.org, 2015).

The Case for Consumers

The ICT-enabled technologies outlined in the SMARTer2030 report could authentically transform people's lives for the better. The report also foresees that over 2.5 billion additional people will be connected to broadband Internet by 2030. A total of 1.6 billion people will have access to eHealth-care solutions and close to a half billion people will study online through on-demand e-learning courses (SMARTer2030.gesi.org, 2015). Innovation goes hand-in-hand with consumer demand and is driven by changes in customer needs, which are the most important factors for new market opportunities for companies. The rapid development of ICTs and the increased focus by customers on "core competencies," such as the ability to respond to the client and anticipate their needs, are also essential factors for innovation and deep transformation needed to achieve SDGs.

Information and Communication Technology and Sustainable Use of Natural Resources

Challenges, Barriers, and Opportunities

Natural resources necessarily play a central role in promoting sustainable development. Presently, we are experiencing a global natural resource crisis, particularly in water resources. ICTs can help in assessing water supply adequacy, modeling different supply and technology alternatives, and factor in different usage technologies. Subsequently, ICTs are relevant to protecting water quantity, enabling water predictions, water management, and aquatic biodiversity.

The following tables show a series of frameworks for the sustainable use of water resources (Tables 3.1 through 3.5).

Tongia et al. (2005) disclosed the importance of ICT's assessments of water supply adequacy, modeling different supply and technology alternatives, and factoring in different usage technologies. Furthermore, the development of dynamic Geographic Information Systems (GIS) for identifying water availability, storage, transmission and distribution, monitoring of water quality, optimization of the allocation between different water uses (e.g., treated drinking water, agriculture, etc.) and water use management at a societal level, including distribution systems (loss reduction) and utilization efficiency, can be applied effectively with the participation of stakeholders. Accessibility of information for decision-making can also support the strategy and provide a scope of different models and solutions (de Jong, 2009).

TABLE 3.1

Water Resources' Challenges

Challenges
1. Provide drinking water to the world's population. This includes the problem of local access nearby, if not in home
2. Provide improved sanitation access to the world's population.
3. Ensure water quality and health standards are met for water consumption.
4. Ensure sustainability of water supplies, e.g., without depleting groundwater resources. This might include technologies for reusing and recycling water for different uses
5. Make water available for non-drinking uses, primarily agriculture, but also commercial and other economic uses.
6. Improve the efficiency of utilization for non-drinking uses such as agriculture, which accounts for the majority of the consumption.
7. Reduce water losses and improve tariff collection.

Source: Tongia et al. (2005).

TABLE 3.2

Water Resources' Barriers

Barriers
1. Water is overwhelmingly subsidized, even in developed countries. The average cost recovery worldwide is estimated to be around 30%. Poor pricing signals can lead to wasteful usage and over usage. The poor often lack public supply of water, and pay a heavy burden for water gathering
2. Lack of accountability and poor decision-making by public officials. Ignorance of mid- and long-term consequences of decision-making; the short-term view overwhelming long-term planning and investment. Unavailability of data and non-transparency in decision-making
3. The linkages between water, agriculture, health care, energy, and economic growth are not well articulated, especially from a planning perspective
4. Planning for water must correlate to the resource base, i.e., the micro and macro watersheds. However, most decision-making and even data collection is based on political or other artificial boundaries and, consequently, decisions are not based on sustainable supply
5. Lack of data on water uses users, alternative supplies, etc., with a temporal and spatial granularity needed for optimal decision-making
6. A system that allows the elite to seek exit strategies that do not scale, e.g., through individual filtering units, tanker supplied water, individual tube wells, etc.

Source: Tongia et al. (2005).

TABLE 3.3

Role of Information and Communication Technology

1. Reducing the number of persons who lack water and sanitation, especially with reliable data of sufficiently detailed granularity (household, rural/urban, regional, etc.)
2. Defined and achieved metrics on local access to water and sanitation – whether in home or within a 5-minute walk, etc.
3. Measured improvements in the quality of drinking and discharged water
4. Stabilization or rise in water tables
5. Publicly available data on water resources, reserves, and their quality for local users, who are empowered to seek redress or other interventions as required
6. Improvement in soil conditions for agriculture, especially related to salinity, chemicals, and other issues dependent on water

Source: Tongia et al. (2005).

GIS is an essential tool for fixing physical dimensions of environmental damage. It is used to automate processes, transform environmental management operations by collecting new information, and support resolutions for environmental stewardship with actions leading to the use and protection of the natural environment through conservation and sustainable practices. The functionality and applications of GIS simplify and automated procedures within environmental management operations. These efficiencies result in significant time savings and help to make intelligent decisions (GIS Solutions for Environmental Management, 2005).

TABLE 3.4

Information and Communication Technology Potential Opportunities

Information and Communication Technology Areas of Opportunity
1. Assess supply adequacy, modeling different supply and technology alternatives, and factor in different usage technologies. This can include the development of dynamic Geographic Information Systems (GIS) for identifying water availability, storage, transmission, and distribution
2. Quality monitoring, especially through low-cost sensors
3. Quality of water impacts health care, agriculture, and industry
4. Optimize the allocation between different uses of water (e.g., treated drinking water, water for industrial usage, agriculture, etc.) via market and non-market mechanisms
5. Water use management at a societal level, including distribution systems – which incorporates loss reduction, equity, etc. – and utilization efficiency
6. To make a meaningful impact, stakeholders must have access to information for informed decision-making, and they must have open access to range of different models and solutions.
7. Help with education regarding efficiency, loss reduction, and new technologies. Reducing losses is especially important for expanding water coverage and availability.
8. Many large developing-country cities only provide water supply for a few hours per day, and 25–50% of the water remains unaccounted for (either lost through poor infrastructure or pilferage).

Source: From Tongia et al. (2005).

TABLE 3.5

Information and Communication Technology Research Opportunities

Examples of Needed Research

1. Low-cost approaches to quality assessment and modeling, including:
  a. Sensors
  b. Data collection (including ad-hoc networks) and sensor integration
  c. Analysis
  d. Dissemination

2. Systems analysis of supply adequacy across a range of uses, technologies, etc., which:
  a. Demands adequate data (e.g., GIS (Geographic Information Systems) based, point-of-use data entry, etc.)
  b. Requires flexible and robust models

Source: Tongia et al. (2005).

Data Management

The science of data management is called "informatics" which is the science of processing data for storage and retrieval, and when it is used for environmental sciences it is called "ecoinformatics" (Oxford Dictionary, 2018). Green Informatics constitutes a new term in the science of information that describes the utilization of informatics in the interest of the natural environment and the natural resources regarding sustainability and sustainable development (Zacharoula S. Andreopoulou, 2009). Using Green Informatics tools, services, and technologies can support actions toward sustainability.

Effective data management has many benefits that translate into improving market position due to transparency, increasing access to capital, and improving decision-making capabilities. It all centers on the data and its usefulness to have comprehensive, accurate, and timely data needed to meet these objectives. A key to good data management is interoperability. It is necessary to combine data from different sources in order to position the process to better understand trends and causal relationships. Doing this requires the development of standards and protocols for describing phenomena, as well as quality control to ensure that the knowledge that results is based on facts. Furthermore, a tool for curating data that can be located in space and time using maps is Data Basin. It provides a way to host and manipulate data sets and create knowledge – to tell a story based upon observations. This science-based mapping and analysis platform that supports learning, research, and sustainable environmental stewardship also provides ranking data sets and social networking tools that interpret, critique, and collaborate in the development of knowledge products to perform more effective actions (Zacharoula S. Andreopoulou, 2009).

The understanding of relations and causalities that are important for improving the development effectiveness of interventions, building better business processes, and/or predicting the outcomes of business models to maximize long-term value have to be tailored to the strategy in place. A vision on how to implement and plan ICT strategically is needed for sustainable development help, overcoming obstacles for adoption of environmentally conscious initiatives. The Strategic Adaptive Cognitive (SAC) framework explicates in a comprehensive method the strategic planning of developing ICTs for sustainable development and the role of leaders in implementing the vision for a transformative ICT (see Chapter 9 "Mechanism Design, Risk Mechanism Theory and Its Relation to ICTs").

Information and Communication Technology for Biodiversity

Ecosystem-based management is an environmental management approach that recognizes the full array of interactions within an ecosystem, including humans, rather than considering single issues, species, or ecosystem services in isolation (Christensen et al., 1996; McLeod et al., 2005). It is necessary to create a strategy for the integrated management of natural resources that are most needed in order to reset ecological balances, to change consumption and production patterns, to promote ecological efficiency, and to restore social equity conditions. Hence, land, water, air, and living resources have to be under a conservation path and sustainable use in an equitable way. The recognition that humans are, with their cultural diversity, an integral component of ecosystems is the first step to develop frameworks and principles of environmental action strategy for sustainable development.

The strategic approach should follow a framework of actions based on the application of appropriate scientific methodologies focused on levels of biological organization which encompass the essential processes, functions, and interactions among organisms and their environment. The use and promotion of ICTs as an instrument for environmental protection and the sustainable use of natural resources is critical considering the accelerated consumption patterns, population growth, and high urbanization rate. ICTs are used for biodiversity preservation, collecting, analyzing, and evaluating a large amount of information in addition to optimizing human behaviors, working processes, and social systems. It can be used in a wide range of applications, delivering complex analysis of diverse information as big data based on the combined knowledge of people, creating new values, efficiently collecting information, and contributing to the avoidance and decrease of biodiversity loss, promotion of sustainable use, and maintenance/extension of biodiversity (Maezawa et al., 2014).

ICT for Biodiversity Preservation:

1. Information Collection
   - Remote sensing of organisms, temperature, and humidity; identification of species by image analysis of organisms, collection of organism information and environment information by using mobile terminals

2. Analysis/Evaluation
   - Evaluation of impacts on organisms, ecosystems, and habitats

3. Information Management
   - Organism information (species, population, habitats etc.), a database for genetic information, etc.

4. Monitoring
   - Monitoring and observation of environment changes and organism behaviors

5. Education, Propagation, and Enlightenment
   - Propagation of information and enlightenment of entire society through network communication technologies and image distribution technologies Further, it is possible to contribute to biodiversity preservation by supporting economic activities, environmental considerations and productivity enhancement in the primary industries (agriculture, fishery, forestry, etc.), which directly involve the supply services that constitute the ecosystem services (Maesawa et al., 2014).

For the preservation of biodiversity and its sustainable use, it is critical to run a PDCA (plan-do-check-act) cycle. This framework is an iterative four-step management method used in business for the control and continual improvement of processes and products. It can also be used to understand the weaknesses of their measures and make necessary improvements to the process, which is essential for companies to make progress in biodiversity conservation. More on PDCA is formulated in the book section titled: "Sustainable Development Goals, PDCA Cycle, and Risk Management Frameworks."

Eco-Industrial Applications and Information and Communication Technology for Industrial Ecology

Many emerging fields striving in sustainability have emerged in recent years mainly with the accelerated progress of ICTs and its applications for environmental science and economic and social change across all areas of human activity worldwide. The object of industrial ecology (IE) is to study industrial systems and their fundamental linkage with natural ecosystems, with the aim to contribute to a more sustainable future (Isenmann and Chernykh, 2009).

Industrial ecology seeks to understand systems using a wide perspective ranging from the scale of molecules to that of the planet. It is about "things connected to other things," a "systems-based, multidisciplinary discourse that seeks to understand emergent behavior of complex integrated human/ natural systems" that can be thought to be composed of interacting technical and social networks integrated into the biosphere (Allenby, 2006; Dijkema and Basson, 2009).

Industrial ecology (where ecology is the science of ecosystems) needs to have a broad and rigorous conceptual framework to approach the long-term evolution of the industrial and economic system. This approach is the fundamental base for the framework to interact with the biosphere and mold an operation strategy that takes into account the benefits and also the detrimental impacts of the action taken. Strategic frameworks present risks and opportunities for improvement within a collective and cooperative plan of action focusing on achieving a well-developed industrial ecosystem with an integrated model of economic activity.

Information and Communication Technology in Agriculture

Developed and developing countries are experiencing a new model of agricultural development where rural areas are expanding in new directions. Traditional societies are being transformed into knowledge societies and old ways of broad basin agricultural activities are being challenged by the new technological era. ICT plays an important role in accelerating the path to developing efficient and sustainable farming systems, such as having location-specific modules of research and extension and promoting market extension, sustainable agricultural development, participatory research, etc., which can potentially enable stakeholders to attain the United Nations SDGs (Meera et al., 2004).

For small-scale farmers, education and training via video, especially when combined with participatory processes, has shown a positive impact on agricultural training and productivity. The dissemination of information for better practices and peer-to-peer learning shows useful methods to improve their activities and share essential information. Agriculture is a source of livelihood for 86% of rural people. Furthermore, agriculture provides 1.3 billion jobs for small-scale farmers and landless workers. According to the International Fund for Agricultural Development (IFAD), GDP growth generated by agriculture is 2–4 times more effective than growth in other sectors. Rural poverty is an issue that needs to be addressed and agricultural activities play a crucial role in reducing poverty and malnutrition. More than 80% of the decline in worldwide rural poverty from 1993 to 2002 was thanks to agriculture. Although the urban population is increasing exponentially, creating new megacities and the 'urban poor effect' around the world, the majority of poor people will live in rural areas until 2040 (Samii, 2008).

The following shows the convergence of applications of IT in support of agricultural and rural development:

- Economic development of agricultural producers;
- Community development;
- Research and education;
- Small and medium enterprise development; and
- Media networks (Meera et al., 2004).

Some agricultural development services that can be provided in the developing world using ICT are:

- Online services for information, education and training, monitoring and consultation, diagnosis and monitoring, and transaction and processing;
- E-commerce for direct linkages between local producers, traders, retailers, and suppliers;
- The facilitation of interaction among researchers, extension (knowledge) workers, and farmers;
- Question-and-answer services where experts respond to queries on specialized subjects; ICT services to block- and district-level developmental officials for greater efficiency in delivering services for overall agricultural development;
- Up-to-date information supplied to farmers as early as possible about subjects such as packages of practices, market information, weather forecasting, input supplies, credit availability, etc.;

- Creation of databases with details of the resources of local villages and villagers, site-specific information systems, expert systems, etc.;
- Provision of early warning systems about disease/pest problems, information regarding rural development programs and crop insurances, post-harvest technology, etc.;
- Facilitation of land records and online registration services;
- Improved marketing of milk and milk products;
- Services providing information to farmers regarding farm business and management;
- Increased efficiency and productivity of cooperative societies through the computer communication network and the latest database technology;
- Tele-education for farmers;
- Websites established by agricultural research institutes, making the latest information available to extension (knowledge) workers and obtaining their feedback (Richardson, 1996).

Information and Communication Technology for Landscape Ecology

The term landscape ecology was introduced by the German biographer Carl Troll (Troll, 1939). Subsequently, several definitions of landscape ecology were published. The main aspects of them can be formulated as the study of:

1. spatial relationships among landscape elements, or ecosystems,
2. the flow of energy, material nutrients, and species among the elements, and
3. the ecological dynamics of landscape mosaics trough time (Roland Grillmayer, 2002).

ICTs and specific software are useful to elaborate data, collect information, and develop different scenarios helping policy-makers to build informed decisions. Therefore, ICTs enhance the adaptation skills of different stakeholders such as policy-makers, public and private sectors, institutions, and final users. Urbanization, population growth, water scarcity for consumption, irrigation and industrial sector use, climatic events, and variations in the ecological characteristics of the region are some of the challenges we face that calls for the need of efficient use of ICTs. Consequently, the use of information systems as a Decision Support System (DSS) will be a key element to develop adaptation and resilience capacities. The DSS is broadly defined as a computer-based system used to aid decision-makers, using data and models to solve unstructured problems (Sprague and Carlson, 1982). Rising food demand and depletion of natural resources are some of the most challenging issues of the twenty-first century. Effective and resource-efficient management of agricultural and environmental systems through the application of ICTs and scientifically sound and knowledge-based information can help address the issues faced by agriculture, landscape, and environmental systems. The DSS is designed to support land managers with complex decision-making and provide tools for priority setting and for measuring changes in the landscape pattern, linkages, fragmentation, diversity. The DSS is also designed to assist in the ability of an ecosystem to support and maintain ecological processes and a diverse community of organisms (Allen et al., 2017) (Figure 3.2).

There are different opinions in terms of the structure of the DSS. The typical DSS consists of three subsystems: the data management, model management, and user interface. The DSS is configured with the four subsystems:

1. the dialog generation and management system (DGMS);
2. the database management system (DBMS);
3. the model base management system (MBMS);
4. the knowledge base management system (KBMS) (Zhang et al., 2009).

A significant component of the DSS is the decision-maker or user and his tasks. Therefore, it can be concluded that such composition of the DSS is the most rational. The DSS is planned to increase the effectiveness of analysis because it offers support for complex problems in the decision-making process and builds efficient information systems (IS) providing the integration

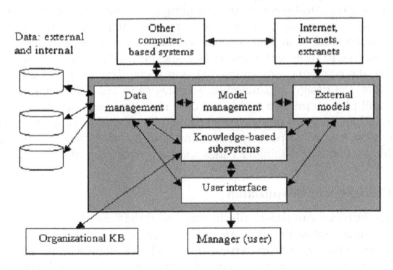

FIGURE 3.2
Schematic view of traditional DSS. (Source from Turban et al., 2005.)

of these systems with an organization's other systems. Effectiveness in the use of DSSs comes with an open structure to adapt to the dynamic environment (Zhang et al., 2009; Decision Support System [DSS] for Public Policy Making, 2014).

Information and Communication Technology for Sustainable Urban Development

The Paradigm Shift: Rural to Urban

For the first time in human history, more people live in cities than in rural areas. Moreover, this increased growth in population moving to cities is leading to the creation of megacities. Megacities have grown fast in the last 20 years; many of them are in developing countries, and their high population density and dependence on a complex infrastructure makes them particularly vulnerable to environmental degradation, climatic events, public health risks, and the effects of urbanization on poverty at the urban level, just to name a few. These transformations suggest a need to rethink the way they operate and adapt to their evolving environments (Marolla, 2017). Cities offer opportunities for economic and social well-being, but realizing the potential of cities will require the adoption of multi-sectoral, multi-stakeholder, and multi-level approaches to sustainable urban development.

The ICT's support strategies for addressing the inequalities between the urban and poor rural areas are by spreading useful information and delivering the tools necessary to achieve a sustainable lifestyle. Science, technology, and innovation can help achieve sustainable urban development when the strategies to cope with those risks associated with urbanization are tailored to the economic, environmental, and social dimensions of urbanism. Therefore, the design and innovative strategies to build sustainable smart cities will play an important role not only for sustainable development but also to achieve several of the current Millennium Development Goals, especially those related to poverty, education, and health (Science, Technology and Innovation for Sustainable Cities and Peri-urban Communities, 2013).

The Challenges of Urbanization

Disease burdens can drastically increase as floods, heat waves, and other severe climate events hit urban areas. Cities are experiencing a higher rate of urban population growth than rural areas, and that, along with the increased percentage of low-income residents that are not capable of mitigating the effects of drastic weather events, magnifies the vulnerability of the urban poor (Parry et al., 2008; Marolla, 2017).

The United Nations conducted a comprehensive study on urbanization and it concluded that the urban population increased from 220 million in 1900 to 732 million in 1950, 3.2 billion in 2005 (World urbanization prospects, 2012) and is estimated to have reached nearly 7.6 billion as of mid-2017 (Table 3.6), implying that the world has added approximately one billion inhabitants over the last twelve years. Sixty percent of the world's people live in Asia (4.5 billion), 17% in Africa (1.3 billion), 10% in Europe (742 million), 9% in Latin America and the Caribbean (646 million), and the remaining 6% in Northern America (361 million) and Oceania (41 million). China (1.4 billion) and India (1.3 billion) remain the two most populous countries of the world, comprising 19% and 18% of the global total, respectively (United Nations, Department of Economic and Social Affairs, 2017).

Major cities in developing countries are experiencing a sharp increase in urban populations compared to industrialized countries. The urban population was comprised of 74% of developed regions and 43% of developing regions. By 2030, approximately 60% of the global population will be living in urban areas. Slums in developing countries help increase the risk of climate change-related impacts, because population density plays a major factor in exacerbating the issues faced by the urban population. The frequency and intensity of heat waves will grow as higher temperatures and the heat-island effect work synergistically in megacities (Parry et al., 2008). Slums and precarious settlements are a tangible reality in urban areas that show deprivation and exclusion among the urban poor. Therefore, all these inequalities and factors shape this particular segment of society and expose vulnerabilities to climate change, which can be in the world's most life-threatening environments (Bartlett et al., 2012; Marolla, 2017).

ICT for cities have a place for the improvement of functions in the urban context because ICTs can be applied to a series of cross-sectoral urban problems. Some of the useful tools of ICTs are geospatial tools for spatial planning,

TABLE 3.6

Population of the World and Regions, 2017, 2030, 2050, and 2100, According to the Medium-Variant Projection

	Population (millions)			
Region	**2017**	**2030**	**2050**	**2100**
World	7550	8551	9772	11184
Africa	1256	1704	2528	4468
Asia	4504	4947	5257	4780
Europe	742	739	716	653
Latin America and the Caribbean	646	718	780	712
Northern America	361	395	435	499
Oceania	41	48	57	72

Source: United Nations, Department of Economic and Social Affairs (2017).

simulation and visualization modeling, mobility tools, solutions for optimizing energy and water management, disaster monitoring and response, and social inclusion.

Geospatial tools for spatial planning

Some of the geospatial tools for use in the urban context are satellite maps and data layers of geographic information systems (GIS):

1. Mapping underground utilities, mines, tunnels and other city infrastructure to identify issues, improve efficiency, and design extensions;

2. Mapping areas at risk of earthquakes, floods, landslides, and other natural disasters, and adjusting development plans;

3. Identifying infill areas such as abandoned land or buildings that are suitable for redevelopment and planning for their reallocation;

4. Mapping natural resources such as prime agricultural land and unique or endangered habitats;

5. Mapping historic and cultural sites that should be protected and designing future urban development that is in cohesion with a city's cultural heritage;

6. Providing virtual addresses to houses and business enterprises that lack formal addresses;

7. Combining multilayer statistical information with satellite maps to run analyses, for example, poverty targeting, urban infrastructure and transport planning, and socio-economic analysis such as crime statistics and tracking illegal settlements (UNCTAD, 2012).

Sustainability initiatives incorporating ICTs into city services and operations require infrastructure changes, but for the most part there is no need for major construction modifications. It is possible to overlap an existing city infrastructure into ICTs tools that improve the city's service demands and operations and allows social linkage, new services without reallocation of scarce resources, or engage in difficult political negotiations. The most effective transformation of a city's operations using ICTs is in the transportation sector. The availability of time travel and mobile phones is a progressive step to accessibility and better traffic information and new traffic services with alternative driving patterns, reducing environmental impacts, reducing time-length driving, and making urban travel enjoyable. The vision of "open source bus" (Mitchell and Casalegno, 2008) expands the idea of commuting as a work/balance tool. In essence, commuter vehicles could be coordinated to aggregate regular users' personal errands, such as dry-cleaning and package delivery, in addition to supporting group study or exercising on longer time travels; all organized into ICT connectivity. Public transit is another important sector where ICTs are re-envisioned as a multi-use service (Spinak and Casalegno, 2012).

Mobility Architecture Programs

Connected Public Transit (CPT) is a set of information services aiming at enhancing customers' experience through ubiquitous connectivity, making it convenient, comfortable, efficient, and reliable. Utilizing real-time information, CPT is an important element in urban transportation and can incorporate various "smart traveler" features that provide dynamic (changeable) guidance. Some CPT features integrate with Personal Travel Assistant (PTA) services that use mobile devices and public monitors located at transit stops and on transit vehicles allowing users to access information during the commute. The Connected Bus, a proof of concept within the CPT program and a pilot project that began in 2007 as part of the Connected Urban Development program at Cisco, is a partnership with the city of San Francisco. Furthermore, reducing the environmental footprint, creating awareness and use of public transportation, and changing the perception of public transportation are among some of the benefits. The following key features display the sustainable scalability of the Connected Bus services:

- Reduction in CO_2 emission – 95% emissions-free hybrid vehicles;
- High-speed wireless Internet access for all passengers;
- Real-time travel information (location, routes, wait times, and more) via on-board touch screens information on the environmental impact;
- Increase operational effectiveness through on-board systems integration, and decrease upgrade costs;
- Improve reliability – this is measured by schedule accuracy, operator and vehicle reliability, supervisor coverage, and congestion management, scaled technologies, public awareness, and an increase of public transportation use;
- Increase effectiveness of the transit operator by enabling efficient communications, relieving the burden of information sharing between driver/rider; An enhanced transit experience, combined with operational efficiencies and "green" benefits derived from a unique configuration of technologies, positions (Mitchell and Wagener, 2010).

Information and Communication Technologies for Disaster Monitoring and Response

Natural and man-made disasters keep mounting in many parts of the world. Risk management serves to identify potential opportunities, and then manage and take action to prevent adverse effects. It also emphasizes the probability of events and their consequences, which are measurable both qualitatively and quantitatively. A risk management framework addresses the full spectrum of challenges in areas such as planning, strategy, operations, finance, and governance. It also recognizes the specific needs of the government's operating model and functions as well as the potential impacts of parallel threats and events. It is important to strengthen governance frameworks and policies and reassert governance roles, establish board-level risk committees with clear roles of the responsibilities of other board committees, and appointed chief risk officers (CROs) that work with Chief Technology Officers (CTOs) to enhance governance frameworks. In essence, ICTs tailor a well-planned risk management framework to cope with uncertainty and reassure that the ICT tools are fully utilized.

The systematic analysis and management of disasters through a well-planned strategic approach to integrating recovery measures, preventing and mitigating risks, and identifying a population's vulnerabilities are a priority to deter or adapt to the aforementioned disaster impacts (Marolla, 2017). Since even the best risk management cannot actually prevent major loss events, the focus must be on managing them. Along with a well-planned strategy and risk management framework, ICTs can improve resilience against natural and man-made hazards. The ICT-based hazard monitoring and surveillance techniques can be used for early warning and land-use planning. ICT plays a critical role in the business continuity process. We can refer to this process as the actions taken in the aftermath of an event and the recovery process that follows. There are frameworks like the business continuity management system (ISO 22301) that supports the process of recovery after a disaster and facilitates the reconstruction and coordination of those displaced from their original homes and communities. The ICTs made the process of recovery more efficient with the use of different tools that included resource management and tracking, communication under emergency situations (e.g., use of Internet communications), collecting essential items for the victims, and national and international fundraising.

The disaster management strategy has no fixed systems in place. It must be flexible and adaptable to the situation, environment, and the socio-economic and geographical area The strategy must maximize the use of its resources to the overall benefit of the community and minimize, in a cost-effective manner, the potential for loss, damage, and injury, while satisfying legal obligations. However, while approaches vary, disaster management strategies should be carried out in a cycle with a continual improvement process.

Science and technology working in conjunction can actually reduce the impacts of risks. Radio science has an integral role in disaster and mitigation of risks. Radio is a crucial tool to monitor the environment and the data are used to predict models that are main factors for safety and economic welfare (Wilkinson and Cole, 2010). Radio communication media is used in disaster warning and management disaster phases. Major radio communication services involve:

- Prediction and detection meteorological services (meteorological aids and meteorological satellite service);
- Earth exploration satellite service;
- Predicting weather and climate;

- Detecting and tracking earthquakes, forest fires, hurricanes, oil leaks, tsunamis, typhoons, etc.;
- Providing warning information by alerting amateur radio services;
- Receiving and distributing alert messages;
- Broadcasting services, terrestrial and satellite (radio, television, etc.);
- Disseminating alert messages and advice to large sections of the public;
- Fixed services, terrestrial and satellite;
- Delivering alert messages and instructions to telecommunication centers for further dissemination to the public;
- Mobile services (terrestrial, satellite, maritime, etc.);
- Distributing alert messages and advice to individuals;
- Relief Amateur radio services;
- Assisting in organizing relief operations in affected areas (especially when other services are still not operational) broadcasting services, terrestrial and satellite (radio, television, etc.);
- Coordinating relief activities by disseminating information from relief planning teams to population;
- Earth exploration satellite service;
- Assessing damage and providing information for planning relief activities;
- Fixed services, terrestrial and satellite;
- Exchanging information between different teams/groups for planning and coordination of relief activities;
- Mobile services (terrestrial, satellite, maritime, etc.);
- Exchanging information between individuals and/or groups of people involved in relief activities (Emergency Radiocommunications, 2017).

Adaptation and disaster risk reduction (DRR) strategies need to be all inclusive and complement each other, focusing on overall risks, development conditions, and local area performance (Brugmann, 2012). From an environmental science approach, where adaptation strategies are emerging with shifting environmental conditions, DRR strategies can extend to social, physical, and economic factors and complement adaptation strategies. Disaster risk reduction strategies are able to identify the wider constraints that determine vulnerability while adaptation strategies focus on developing hazard forecasting and early warning systems. Theoretically, DRR extends beyond disaster preparedness measures exclusively (Venton and La Trobe, 2008). Early Warning Systems (EWS) for floods, for example, are in place in a wide range of countries and are based on the Internet and sensors network to monitor data from various deployed sensor modes to capture data. Flood forecasting and early warning is one of the most effective flood risk management strategies to minimize the negative impacts of floods (United Nations Economic and Social Commission for Asia and the Pacific, 2015). Monitoring and forecasting flood disaster and developing a risk assessment and business continuity management system to better prepare for responses to disasters are necessary for planning and policy-making and to adapt to the vulnerabilities of severe weather impacts (Shafiq et al., 2014).

Cities' governance and preparedness are an integrated part of the strategy and ICTs that are used for "dashboards" or operations centers combine data from different departments allowing monitoring risks and analyses of data from sensors that are established throughout the city to detect and resolve critical infrastructure and safety issues (ICT for Data Collection and Monitoring & Evaluation, 2013).

Below are notable examples of applying ICTs for disaster resilience in cities:

1. Rio de Janeiro (Brazil) set up an operations center that displays real-time integrated data from 30 agencies, which helped improve coordination and reaction times;

2. Mumbai (India) has 35 automatic weather stations that measure real-time rainfall intensity and flow gauges on the Mithi River to monitor water flow;

3. Chacao (Bolivarian Republic of Venezuela) has a wireless early warning system that connects civil protection and environmental institutions with cameras that monitor four river channels crossing the city and shares online real-time hazard information with citizens (Science, Technology and Innovation for Sustainable Cities and Peri-urban Communities, 2013).

The Link between Natural Resources and Economic Growth

Information and Communication Technology and CO_2 Emissions Reduction

Economic growth, population well-being, and a sustainable future are inextricably linked. Nonetheless, economic growth is not the same as economic development. Development alleviates people from low standards of living into proper employment with suitable shelter. Economic growth, which is a phenomenon of market productivity, does not necessarily lead to the preservation of natural resources, economic, social and environmental well-being of populations and ecosystems. In that context, SDGs must be aligned with sustainability initiatives and meeting the needs of the present without compromising future generation's necessities. The economic growth of countries, particularly in the developing world, is leading to rising CO_2 emissions levels. Population growth, rapid urbanization, and industrialization, as well as the growth of industries such as ICT and tourism, are some of the challenges developed and developing countries are facing to come up with solutions to efficiently cope with the deterioration of natural resources, health risks due to air pollution, and other health hazards. Moreover, international and global inequalities, such as global North-South differences in historic CO_2 emissions (Chancel and Piketty, 2015; Jorgenson,2014; Rosa and Dietz, 2012), disproportionate impacts of climate effects (IPCC, 2014; Roberts and Parks, 2006) and power imbalances between nations in the global North and South with respect to climate policy (Ciplet et al., 2015; Dunlap and Brulle, 2015) are also crucial issues that need urgent attention in order to reach a sustainable development plan where societies can dramatically improve their quality of life, the environment, and governance. In regards of industries tackling carbon emissions while pursuing economic growth, there is a big disparity throughout industries between rapid- and moderate-following industries. The aforementioned issues among countries and industries increases the importance of the ICT sector in finding new tangible solutions. It also raises new questions for governments regarding the implementation of policies and initiatives to promote economic, social, and environmental well-being and create policy frameworks to reduce greenhouse gas emissions (Jorgenson et al., 2017).

Technology can play an increasingly significant role in helping reduce carbon emissions and the role of ICT industries in reducing emissions can have global implications. Given its high energy demand, the ICT sector is still a net source of GHGs. Data centers and other sources are contributing to GHGs and, in the case of data centers, it contributes approximately 2% of global GHG emissions – on par with the aviation sector. However, according to the Global e-Sustainability Initiative (GeSI), ICT has the potential to cut 20% of GHG emissions by 2030 by helping companies and consumers to use and save energy more intelligently.

Within companies, there are several ways for reduction strategies. The categories of emission levels are subdivided by three opportunities. The following shows how the ICT sector needs to consider the various categories:

- Scope 1 covers the GHG emissions generated by facilities within the boundaries of an organization. For most ICT companies, this usually refers to the combustion of fuels for heating offices, the power supply of ICT equipment either for backup or, in more recent times, for cogeneration, or the power needed to cool ICT equipment. It also covers emissions resulting from intentional or unintentional releases of coolant from equipment, such as air conditioning plants in data centers. Scope 1 emissions will also cover emissions from vehicles owned by ICT companies.

- Scope 2 covers the indirect emissions from the generation of pur-
chased electricity, heat, or steam consumed by the organization.
Most ICT organizations buy the majority of their electricity from
utilities in order to power their computers, data centers, communications equipment, heating, lighting, cooling, and use of other office
equipment. As a result, most emissions from the internal operations
of ICT companies are reported in this category.
- Scope 3 covers a company's entire value chain emissions impact and
enables a company to track the full impact of its upstream and down-
stream impacts. It is at the company's discretion what it reports under
scope 3 and most companies will keep it to easily measured items
such as corporate business travel. Full life-cycle scope 3 reporting is
still very rare and it is by far the largest component of most organizations' carbon footprint (Toolkit on Environmental Sustainability
for the ICT Sector, 2012; The Potential Global CO_2 Reductions from
ICT Use, 2008) (Graph 3.3).

Ericsson and Columbia University Earth Institute developed a report titled:
"How Information and Communications Technology can Accelerate Action
on the Sustainable Development Goals" (2016) presenting the potential role
of ICTs and its role in SDGs in creating the transformation of societies to

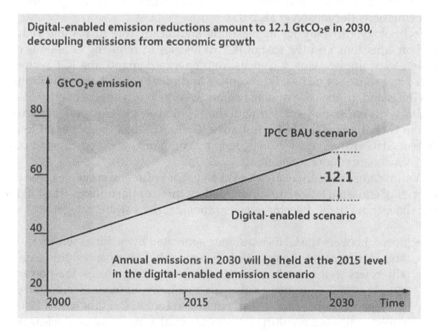

GRAPH 3.3

Digital-enabled CO_2 emissions trajectory toward 2030, compared to IPCC BAU scenario. (From
GeSI, #SystemTransformation report, 2016.)

appoint pragmatic solutions to the challenges of the twenty-first century involving the solicitation of technological solutions (ICT & SDGs Final Report, 2016). Major progress has been made achieving the SDGs and how ICTs can help face the challenges, demonstrating the scale and ambition of its agenda. In summary, ICTs objectives are:

- Accelerated upscaling of critical services in health education, financial services, smart agriculture, and low-carbon energy systems;
- Reduced deployment costs, addressing urban and rural realities;
- Enhanced public awareness and engagement;
- Innovation, connectivity, productivity, and efficiency across many sectors;
- Faster upgrading in the quality of services and jobs.

The accelerated impact of ICTs in society brings new issues that must be considered. The risk and opportunities are greater than ever and by focusing on positive consequences of a risk the elements of both risks and opportunities should have an ICT context and be directly or indirectly related to the ICT sector's pragmatic solutions. No technology is without risks and ICTs increase a number of issues related to its use and applications that have been identified:

- Privacy and surveillance;
- Cybersecurity;
- Loss of human skills;
- Possible public concern about health effects;
- Electronic waste and carbon emissions;
- Digital exclusion;
- Child protection and the Internet (ICT & SDGs Final Report, 2016).

An Overview of the Role of Information and Communication Technology in Each of the Sustainable Development Goals

The SDGs won't be met without active citizen's participation and the developments and applications of new knowledge. That is the crucial role of ICTs in shaping policies, providing guidance for the research and analysis of solutions in addition to providing information on how to prioritize and create a plan of action on each of the goals. ICTs create possibilities and resources.

Sustainable Development Goal 1: End Poverty in All its Forms

- Timely and accurate information assist in ensuring equal rights to economic resources, improving productivity, and accelerating the well-being of populations, providing a better means of income and livelihood. Disruptions in financial inclusion creates inequality and ICT-enabled solution services such as mobile banking and micro-crediting, giving access to markets for small producers and their products, furnishes a more equal society of financial opportunities and minimizes poverty levels, at the same time creating a more stable economic class (ICT & SDGs Final Report, 2016; Proposed ICT Indicators for the SDG Monitoring Framework, 2015).

Sustainable Development Goal 2: End Hunger, Achieve Food Security and Promote Sustainable Agriculture

- ICTs help to rethink how we grow, share and consume our food by giving farmers direct access to market information, weather forecasts, logistics, and storage to help increase agricultural yield, restore soil, reduce waste, and improve productivity. Sustainable agriculture is one of the biggest challenges to accomplish SDGs. This challenge will become unconquerable without a specific and understandable plan to bring permanent resolutions and with the particular mechanisms for ICT applications, it can broadly follow the ways of improving productivity for a sustainable agriculture and achieve food security, ending hunger (Proposed ICT Indicators for the SDG Monitoring Framework, 2015).

Sustainable Development Goal 3: Ensure Healthy Lives and Promote Society's Well-Being

- The goal by 2030 – to reduce by one third premature mortality from non-communicable diseases through prevention and treatment of substance abuse, including narcotic drug abuse and harmful use of alcohol and promote mental health and well-being – can be strengthened by ICTs that allow health workers to be connected to health information and diagnostic services. A global health-care ecosystem that amalgamates the prevention and treatment of diseases, morbidity and mortality population rates with connectivity and digital technologies that can make projections about diseases and patient knowledge and attitudes, and enable the effective and continuous management of diseases and health practices is primordial to ensure the well-being of the global community.

Sustainable Development Goal 4: Ensure Inclusive and Equitable Quality Education

- Encourage and promote opportunities for all with a quality education is essential for creating prosperity, workers with the right skills integrating into the new economy, knowledge and equitable society. The challenge rest in vulnerable populations, including persons with disabilities, indigenous people, refugee children and poor children in rural areas. The accessibility to education is key for resources and opportunities that provide the tools necessary to achieve a prosperous society even in remote or low-income areas. Teachers can prepare for classes anytime or anywhere. Therefore, ICT opens up access to education to underserved populations developing educational opportunities leading to improved socio-economic opportunities.

Sustainable Development Goal 5: Achieve Gender Equality and Empower all Women and Girls

- Women experience many forms of violence that in many cases end up in death. For example, on the basis of data from 2005 to 2016 for 87 countries, 19% of women between 15 and 49 years of age said they had experienced physical and/or sexual violence by an intimate partner in the 12 months prior to the survey and as stated before, in the most extreme cases, such violence can lead to death. In 2012, almost half of all women who were victims of intentional homicide worldwide were killed by an intimate partner or family member, compared to 6% of male victims. Women and in general any demographic minority facing inequality can gain advantages from ICTs, which support gender equality by leveling the playing field. Connectivity gives them access to information that helps them with their productive, reproductive, and community roles. Moreover, sharing information through mobile networks provides better awareness and knowledge on important issues related to their well-being. ICTs give access regardless of gender to markets, education, training, and employment for greater opportunities in sustainable livelihood (Proposed ICT Indicators for the SDG Monitoring Framework, 2015).

Sustainable Development Goal 6: Ensure Availability and Sustainable Management of Water and Sanitation

- Water, once an abundant natural resource, is becoming a more valuable commodity due to droughts and overuse. Therefore, the management of water resources under set policies and regulations are crucial to making water and sanitation more widely available. ICTs also assist in determining infrastructure location, supporting cost-saving and low-maintenance programs, optimizing operations, and improving quality of service (Proposed ICT Indicators for the SDG Monitoring Framework, 2015).

Sustainable Development Goal 7: Ensure Access to Affordable, Reliable, and Sustainable and Modern Energy for All

- Affordable, reliable, and sustainable energy is essential to achieve SDGs, from poverty eradication via advancements in health, education, water supply and industrialization to mitigating climate change. The rate of energy access varies among developed and developing countries and currently falls short of what will be required to achieve the SDGs. The ICT-based solutions have already been an important tool for organizations to achieve energy efficiency and reduce greenhouse gas emissions. Many ICT industry leaders are also making progress in more environmentally sound and less carbon-intensive products and services. These include smart grids, smart buildings and homes, and smart logistics that improve energy efficiency and lower energy consumption across different sectors (United Nations SDG 7, 2013).

Sustainable Development Goal 8: Sustained, Inclusive, and Sustainable Economic Growth, Employment, and Decent Work

- Sustained and inclusive economic growth provides essential components to deliver benefits such as increasing labor productivity, reducing the unemployment rate especially for young people, and improving access to financial services. In this case, ICTs or digital technologies are transforming day-to-day business in both traditional and new sectors. ICTs support the goal of achieving higher levels of productivity from economies through diversification, technological upgrading, and innovation, focusing on high value-added and labor-intensive sectors. Many forms of employment in the twenty-first century require ICT skills (United Nations SDG 8, 2013).

Sustainable Development Goal 9: Build Resilient Infrastructure, Promote Sustainable and Inclusive Industrialization, and Foster Innovation

- Aging, degraded, or non-existent infrastructures make conducting good business challenging. Materials, resources, labor, and service are essential components of business and the ability to access them efficiently is vital to establishing new markets. Most businesses today rely on computing and technology-based skills and the global economy demands accessibility to international markets and understanding new trends. Conversely, basic infrastructure supporting technologies, communications, transportation, and sanitation is not universally available, detrimentally impacting economic growth, impeding equality and societal progress. ICTs play an essential role, especially for emerging information and knowledge societies. Some contributions of ICT are open access to academic research, transparency to make informed and efficient decisions, and platforms for online collaboration in many phases of learning activities, work, and management initiatives – some form of economic strategy that brings different actors together in order to produce a mutually valued outcome (United Nations SDG 9, 2013).

Sustainable Development Goal 10: Reduce Inequality

- Reducing inequality involves promoting economic inclusion and empowering the bottom percentile of income earners. Income inequality is a global problem that requires global solutions and by increasing access to information and knowledge, ICT helps reduce inequality, assisting policy-makers and global leaders to develop initiatives for social and economic progress, even to disadvantaged segments of society, such as persons with disabilities and marginalized populations (United Nations SDG 10, 2013).

Sustainable Development Goal 11: Make Cities Inclusive, Safe, Resilient, and Sustainable

- As we face global trends of climate change, urbanization of world populations, loss of habitats and ecological connectivity, and rising impacts to people and property due to natural disasters, those responsible for design of the built environment can potentially shape the well-being of future generations in terms of public health (or risk) and economic opportunity (or burden). Indeed, one of the most important tools for human adaptation is the planning and engineering of environmentally sound solutions, incorporating information gained incrementally to anticipate and holistically address ongoing and future needs. Sustainable development has been recognized for decades as the aspirational goal to guide universal action (Marolla, 2017). The ICTs offer cities innovative infrastructure and applications, such as smart buildings, smart water management, intelligent transport systems, and efficiency in energy and waste management system – enabling the allocation of each sorted waste fraction to the proper treatment and recycling processes. These lead to a more effective and holistic city management process. Smart city mobility, mobile ride sharing, e-mobility, driver-less transportation, intermodality, connected infrastructure/Internet of Things (IoT) all help to reduce resource consumption, improve energy efficiency and reduce air pollution which reduces environmental issues, carbon emissions, and health risks. Furthermore, the smart building is another area to achieve resilience and sustainability that incorporates alarm management and automation, big data analytics, energy management, smart metering, as well as IoT/sensors, monitoring, detection and diagnosis technologies, which within the health-IT sector has the potential to improve diagnosis and reduce diagnostic errors. In general, approximately a 5% CO_2e emissions savings in 2030 from smart building and smart city mobility alone can be attained (Ono et al., 2017).

Sustainable Development Goal 12: Sustainable Consumption and Production Patterns

- As defined by the Oslo Symposium in 1994, sustainable consumption and production (SCP) is about "the use of services and related products, which respond to basic needs and bring a better quality of life while minimizing the use of natural resources and toxic materials as well as the emissions of waste and pollutants over the life cycle of the service or product so as not to jeopardize the needs of further generations" (Sustainable Consumption and Production Indicators for the Future SDGs UNEP Discussion Paper, 2015).

- SCP links environmental and social issues with economic activities and markets. It has a holistic approach toward the production and consumption side. Information communication and technologies help in changing practices to cleaner production and the eco-efficiency of production systems. Hence, a producer's responsibility in pollution control legislation, and investments in innovation, green technologies, and resource efficiency accompanied by policies and measures to implement those initiatives, become more viable with ICT innovations and applications. Nonetheless, increased dematerialization and virtualization and smart technologies for sectors such as agriculture, transportation, energy, supply chain management, and smart buildings are main drives for an effective transformation and a circular economy, which is key for sustainable development (Sustainable Consumption and Production Indicators for the Future SDGs UNEP Discussion Paper, 2015).

Sustainable Development Goal 13: Urgent Actions to Combat Climate Change and its Impacts

- A changing climate can alter the pattern of diseases, mortality, human settlements, food, water, and sanitation. Climate change brings increasing temperatures, rising seas, and more frequent incidence of severe storms. These effects produce dangerous sanitary events and are known health risks (Jensen, 2007). In addition, shifting rainfall patterns may influence transmission of many diseases, including water-related illnesses such as diarrhea, and vector-borne infections, including malaria. Climate change could have far-reaching effects on how food is produced and may have health impacts from increasing rates of malnutrition (Marolla, 2017). The Anatomy of a Silent Crisis (Human Impact Report Climate Change: Global Humanitarian Forum, 2009) confirmed that every year more than 300,000 people die of climate change consequences, 325 million people are seriously affected, and economic losses total US$125 billion. Four billion people are vulnerable, and 500 million people are at extreme risk (Annan, 2009). All the aforementioned risks associated with climate change impacts propel nations to take urgent action to combat climate change. ICT can deter and/or minimize risks by enabling:

- *Access*: ICT plays a critical role in collecting and sharing data on climate, weather, information for forecasting weather events, and in early warning systems.
- *Connectivity*: Connectivity to mobile phones, apps, or media can build awareness about climate risks and improve levels of preparation and resiliency. Through connectivity via apps and online platforms, ICT can also help foster awareness and cultural momentum around sustainable consumption and greener lifestyles.
- *Efficiency*: ICT enables multiple "efficiency effects" via technologies such as cloud computing; through smart applications that provide clean solutions for manufacturing, transport systems, and infrastructure; and by helping to identify areas for further efficiency via big data collection and analytics. ICT applications support a system of efficiency for energy, transport and buildings, manufacturing, smart services, and agriculture and urbanization. Smart applications also help in adaptation strategies and foster resilience such as real-time weather information updates (ICT Sustainable Development Goals Benchmark: Connecting the Future, 2017).

Sustainable Development Goal 14: Conserve and Sustainably Use the Oceans, Seas, and Marine Resources

- Ocean sustainability and conserving seas and marine resources is fundamental to our existence. Human actions are putting these ecosystems in jeopardy and despite efforts to battle these detrimental actions, the oceans' health and ecosystems continue to decline.

  Satellites can collect comprehensive data over the oceans, arctic areas, and other sparsely populated zones that are difficult for humans to monitor. It also provides information about the color of the ocean. For example, color data helps researchers determine the impact of floods along the coast, detect river plumes, and locate blooms of harmful algae that can contaminate shellfish and kill other fish and marine mammals. Moreover, one of the most significant potential impacts of climate change is sea level rise, which can cause inundation of coastal areas and islands, shoreline erosion, and destruction of important ecosystems such as wetlands and mangroves. Satellite altimeter radar measurements can be combined with precisely known spacecraft orbits to measure sea levels on a global basis with unprecedented accuracy. Nonetheless, big data can be used to analyze biodiversity, pollution, weather, and ecosystem evolution to help plan mitigation and adaptation strategies (How Are Satellites Used to Observe the Ocean? 2015). In conclusion, ICTs can play a significant role in the conservation and sustainable use of the oceans with efficient monitoring providing accuracy and accountability in its reporting.

Sustainable Development Goal 15: Protect, Restore, and Promote the Sustainable use of Terrestrial Ecosystems, Sustainably Manage Forests; Combat Desertification and Halt and Reverse Land Degradation and Biodiversity Loss

- The depletion of natural resources causes environmental degradation and leads to poverty, inequality, and deprivation, impending populations from achieving social and economic opportunities. ICT can help to achieve SDG 15 in making improved monitoring and reporting for the conservation and sustainable use of land; protect, restore, and promote sustainable use of terrestrial ecosystems; sustainably manage forests, combat desertification, and halt and reverse land degradation and biodiversity loss. This includes the use of big data to analyze short- and long-term trends and to develop strategies for mitigation. Sensors, data collection, and analysis also help in land restoration. Good governance and policies aligned with new technologies and local communities, strategic partners as well as public and private actors' engagement are all significant factors and the ability to combine measures for sustainable use of terrestrial ecosystems.

Sustainable Development Goal 16: Peace, Justice, and Strong Institutions

- Sustainable development is a reflection and goal of many meaningful actions for a society with permanent progress; effective state administrations with responsible institutions, transparency, and the rules of law that have an intrinsic value of their own. Therefore, peace, stability, and protecting human rights represent the basis of good governance. ICT assists in crisis management and peace building through powerful tools such as electoral monitoring. The use of open data increases transparency, empowers citizens, and drives economic growth (Proposed ICT Indicators for the SDG Monitoring Framework, 2015).

Sustainable Development Goal 17: Strengthen the Means of Implementation and Global Partnerships for Development

- Long-term policy attainment and the successful implementation of the SDGs requires a multi-sector participation enabling the exchange of technology, knowledge, experiences, policies, and good practices; for example, mobilizing resources and investments in multiple areas like financing climate action. ICT can assist with the implementation of SDGs by encouraging international cooperation and coordination, promoting technology transfer and capacity building, strengthening multi-stakeholder partnerships, and enabling data monitoring and accountability. ICTs are catalysts that accelerate all three pillars of sustainable development – economic growth, social inclusion, and environmental sustainability.

- Rural and urban planning utilizing ICTs needs to be included in government and policy-maker measures, and support ICT-enabled transformation. Many government processes – payments, tax collection, procurement, training, human resources, program design, public deliberation, information management, analytics, legislative drafting, even voting – should be upgraded with the transformative capability of ICT tools and resources. The new technology advancements in a fully "Network Society" will create new jobs and the way we develop new skills is quickly changing. The ICT must ensure that no one is left behind in the digital era and, in fact, will be included in times of transformation (Proposed ICT Indicators for the SDG Monitoring Framework, 2015).

Sustainable Development Goals, Plan-Do-Check-Act Cycle, and Risk Management Frameworks

Developing Sustainability Standards

The Sustainable Development Goal 15 of the 2030 Agenda for Sustainable Development is devoted to "protect, restore, and promote sustainable use of terrestrial ecosystems, sustainably manage forests, combat desertification, and halt and reverse land degradation and halt biodiversity loss." For the preservation of biodiversity and its sustainable use, it is critical to run a

PDCA (plan-do-check-act) cycle. The PDCA cycle assessment should follow these steps:

- *Plan*: Thereafter, based on the analysis/evaluation results, it is requested to develop preservation and utilization plans to stipulate how the targeted area should be preserved and used, and who takes what action in which time- frame.
- *Do*: Following this stage, actions should be taken for preservation activities and sustainable use based on these preservation and utilization plans.
- *Check*: To understand accurately how many wild animals and plants live and grow in which location inside the targeted area.
- *Act*: Then, it is essential to carry out analysis and evaluation based on this study and monitoring data and to understand the current status and the time-related changes.

Finally, the results are used as feedback for the next implementation plan, which provides continues improvement process, assessing the risks and opportunities, and creating an effective way to cope with risks.

ICT innovations are particularly important for supporting the development processes in PDCA. Quality and quantity of innovation processes include creativity, knowledge capture, transfer, and sharing that can be aligned and tailored to other standard frameworks. The ICTs are useful for collecting data, analyzing, and evaluating large amounts of important information and optimizing human behaviors, social, and work systems in an efficient way. These processes need a continual improvement cycle for efficiency and the conjunction of PDCA cycle (as the dynamic cycle that can be implemented with elements and requirements of a quality management system) and risk management frameworks develop a comprehensive strategy to address SDGs. ICT considers risk management to be a continuous process, a primary part of the activities, which is to include policy and a plan of action in control systems and procedures at every level of the organization and enable measured performance parameters that correlate to its strategy in a consistent and recognized manner (Management of Continual Improvement for Facilities and Activities: A Structured Approach, 2006). After considering the relation between comprehensive quality management systems such as PDCA and ICT, we can focus on the advantages of utilizing ICTs for biodiversity preservation with a risk management framework:

1. *Information collection*: Remote sensing of organisms, temperature, and humidity; identification of species by image analysis of organisms and collection of organism information and environment information by using mobile terminals
2. *Analysis/evaluation*: Evaluation of impacts on organisms, ecosystems and habitats

3. *Information management*: Organism information (species, population, habitats etc.), database for genetic information, etc.

4. *Monitoring*: Monitoring and observation of environment changes and organism behaviors

5. *Education, propagation, and enlightenment*: Propagation of information and enlightenment of entire societies through network communication technologies and image distribution technologies (Maezawa et al., 2014) (Figure 3.3).

Information and Communication Technology Innovation Policies Management and Governance

Inclusive political and economic institutions are important preconditions and requirements for sustainable development. The transparency and quality of national and sub-national governance, which is a major determinant of many variables associated with the well-being of individuals within a country, is designed to encourage high standards of behavior and operations. In addition to the transparency and quality of governance at national and sub-national levels, there is also now greater attention to global governance issues (The OECD Corporate Governance Principles, 2015). In a world where risks transcend borders and sectors, with implications that affect the international financial market, the effects of climate change across nations and regions, – environmental degradation, public health hazards, the spillover from intra-state conflict and impact of international crime, terrorism and illicit financial flows – put cross-border governance issues at the front page emphasizing the importance of *how* a government delivers its policies,

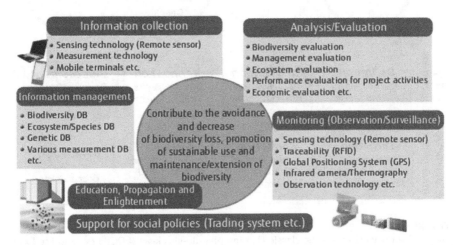

FIGURE 3.3
Possibility of biodiversity preservation using ICT. (From Maezawa et al., 2018.)

instead of *what* a government delivers in its policies – irrespective of their nature and degree or provision – in an effective and impartial way without corruption (Charron et al., 2013).

The importance of the nexus between ICT, PDCA, and governance is attributed to the improvement of the success rate of the national or local government activities. Therefore, monitoring the ICT innovation projects – validating the related policies and evaluating the effective direct and indirect impact on the areas affected – is an effective process to assure efficiency in the capability to continuously improve their services via an objective evaluation of the resulting impact, the local government stakeholders, citizens, and enterprises (Candiello and Cortesi, 2011).

Figure 3.4 shows a comprehensive framework for monitoring the ICT innovation projects; substantiating policies for innovation and assessing the risks and opportunities of the areas targeted, which leads to the potential of improving the management system and government initiatives. This strategic approach is vital to achieving efficiency in delivering initiatives and policies aimed at creating a sustainable development society. The role of the public and private sector and other stakeholders will need to be established with close collaboration and strong partnerships and to seek support from international organizations.

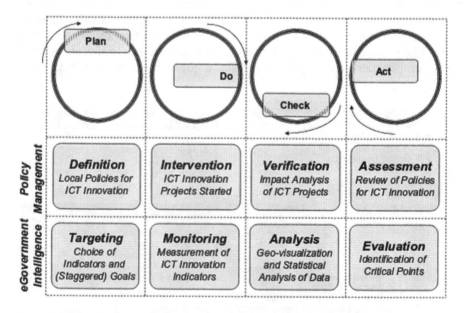

FIGURE 3.4
Adapted frameworks with the classic Deming plan-do-check-act (PDCA) cycle to the local government requirements for ICT innovation policies management. (From Candiello and Cortesi, 2011.)

Information and Communication Technology Risks and Opportunities Related to Key Sustainability Issues

Sustainability has emerged as a competitive advantage for organizations and serve as the basis of achieving SDGs, considering the long-term aspect of the concept of sustainability and its ethical principle of achieving equity between the present and future generations (Diesendorf, 2000).

There are risks that need to be addressed in the use of ICT which is vital for maximizing the sustainability potential of its products and services. Minimizing the detrimental impact and reducing toxic materials as well as energy consumption during production are some of the steps needed to be taken, leading to the "use phase" of the process, emphasizing the value of engaging with both the business and private consumers of ICT products and services (Madden and Weißbrod, 2008). Other hazards of the use of ICTs are carbon emissions. Global consultant Gartner estimates that ICTs presently account for approximately 0.86 metric gigatons of carbon emissions annually, or just about 2% of global carbon emissions (Gartner, 2009). In comparison, the Global e-Sustainability Initiative (GeSI) 2008 report, SMART2020, estimated that the ICT sector's emissions would reach 1.43Gt CO_2e by 2020, which would represent 2.7% of global emissions. Five years later, SMARTer2020 report revised that forecast down to 1.27Gt, representing 2.3% of global emissions. The revised estimates were based on actual energy efficiencies realized between 2008 and 2012 as well as on updated data. Finally, the SMARTer2030 report (2015) predicts a further decrease, with ICT's own footprint expected to reach 1.25Gt CO_2e in 2030, or 1.97% of global emissions (SMARTer2030.gesi.org, 2015).

Although GeSI estimations showed a prediction in the reduction of ICT's own footprint, the urgency of minimizing the environmental impact of ICTs is significant due to the sector's dynamic transformation, the needs of the new "Knowledge Society" and demand of its services.

The International Telecommunication Union (ITU) has estimated the contribution of ICTs (excluding the broadcasting sector) to climate change to be between 2% and 2.5% of total global carbon emissions. The main contributing sectors within the ICT industry include the energy requirements of PCs and monitors (40%); data centers, which contribute a further 23%; and fixed and mobile telecommunications, which contribute 24% of the total emissions (ITU, 2009).

We need to take into consideration the global economy is creating a new middle-class model, particularly in developing countries. Hence, production and consumption of ICTs are increasing exponentially and as a consequence, emissions from the manufacture and use of PCs alone will double over the next decade. Moreover, in the next 15 years, another 1.8 billion people will enter the global consuming class and worldwide consumption will nearly double to US$64 trillion as developing economies will continue to drive economic growth and determine consumption patterns (Manufacturing the Future: The Next Era of Global Growth and Innovation, 2012).

A Contemporary Type of Risk: E-Waste

Methods of disposal, including incineration of electronic waste (e-waste), are another key environmental issue. Among the discarded items are disused mobile phones, obsolete computer and television equipment, old cables, phone chargers, batteries and other ICT hardware. In addition to its hazardous components being processed, e-waste can give rise to a number of toxic by-products likely to affect human health. Furthermore, recycling activities such as the dismantling of electrical equipment may potentially bear an increased risk of injury (Electronic Waste, 2017).

The hazards expand in many countries as disposable trash container sites for domestic waste are used to discard disused mobile phones, old computers, and other electronic junk. When those items are incinerated alongside other solid waste materials it creates carcinogenic emissions. At the same time, the failure of manufacturers to maximize the life span of equipment increases the burden of emissions from the manufacturing sector.

ICT products are resource-intensive in manufacturing and production and consume great amounts of energy while in use. These actions create solid and toxic waste that poses a risk to human health and the environment (Figure 3.5 and Table 3.7).

Generally, in comparison with many other industries, the ICT production sector does not have the same impact and levels of demand on natural resources. However, it is necessary to highlight that the design, manufacture,

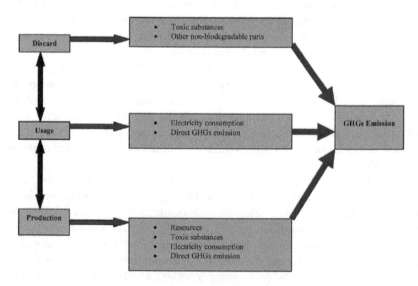

FIGURE 3.5
Adapted from "Emissions GHGs Arising from Three Dimensions by ICT Industry". (From Unhelkar, 2011.)

TABLE 3.7

Environmental Effects and Opportunities Created by Information and Communication Technology

First Order Effects	Second Order Effects	Third Order Effects
Design and manufacture of ICT equipment • ICT production is a relatively lightweight industry • Use of toxic components • New waves of technology are more energy efficient	**Increase and decrease in use of transport** • Increase in home deliveries as a result of E-commerce will have significant environmental impact unless well coordinated • Telework reduces travel miles for employees • Telematics reduces traffic congestion, journey times and therefore pollution • Rebound effects from increased leisure travel	**De-coupling economic growth and energy consumption** • Possibilities of reducing energy used per unit GDP
Operation of ICT equipment • Energy use even in stand-by mode	**ICT in business systems** • B2B E-commerce and ICT-managed control systems create efficiencies and reduce environmental impact	**De-coupling economic growth and carbon emissions** • Possibilities of reducing carbon dioxide emissions per unit GDP
Disposal of ICT equipment • Problematic disposal • Recycling and Safer designs increasing	**Virtualization of material products** • Possible environmental savings from increased trade in intangibles	**Changing settlement patterns** • Conflicting pressures on local settlement • Possible increase in environmental pressure on regions
	Effects on product lifetimes • Some E-commerce business models extend product lifetimes • Product development cycles are often reduced by use of ICT	
	Distribution and manipulation of environmental information • Significantly enhanced by ICT	

Source: Environmental Effects and Opportunities Created by Information and Communication Technology: The Impact of Information and Communication Technology on Sustainable Development (2002).

operation, and disposal of ICT does impact the environment harmfully. The ICT industry can become more sustainable with appropriate policies and implementation of management systems that maximize the efficiency of processes, establish environmental sustainability guidelines for the sector, and play a valuable role in enabling other industries, and in the end, educating the consumer to help accomplish the SDGs.

Guidelines for Environmental Sustainability for the Information and Communication Technology Sector

Investment in technology is a key accelerator to help nations achieve the United Nations SDGs. The ICT benefits for society are well documented but the sector also has challenging and problematical roles for environmental sustainability. Engaging in sustainable practices is a key component of the industry awareness, leading to pragmatic actions to assessing its impacts. The industry has increased sustainability by decreasing resource intensity, but has at the same time encouraged resource consuming lifestyles in part helped by booming economies in the developing world. To achieve SDGs, ICTs needs to make the overall impact clearly environmentally sustainable. The first steps in regards to ICT impacts analysis were understood to be limited to the direct effects of ICT on the environment. The progression of sustainable ICTs drove many elements, such as management systems for continues improvement, supply chain, energy efficiency, etc. Presently, the efforts include the use of ICT to improve the environmental efficiency of other industries and domains, focusing on its effectiveness on mutual understanding and changing the system level activities, i.e., the complex web of behavior of people, institutions, organizations, and political jurisdictions (Thöni and Tjoa, 2015). There are several institutions taking the lead in environmental sustainability actions to drive their own business performance, while being more responsible corporate citizens. Setting standards for better performance, effectiveness, and the maturity of a company's sustainability program is a goal that brings the organizations' objectives and strategies into alignment and enables peer benchmarking and best practice. The International Telecommunication Union (ITU), originally the International Telegraph Union, is a specialized agency of the United Nations that is responsible for issues that concern ICTs. The ITU has established a series of resources and components to accomplish the aforementioned objectives:

- Sustainable ICT in corporate organizations, focusing on the main sustainability issues that companies face in using ICT products and services in their own organizations across four main ICT areas: data centers, desktop infrastructure, broadcasting services, and telecommunications networks.

- Sustainable products, where the aim is to build sustainable products through the use of environmentally conscious design principles and practices, covering development and manufacture, through to end-of-life treatment.

- Sustainable buildings, which focuses on the application of sustainability management to buildings through the stages of construction, lifetime use, and decommissioning, as ICT companies build and operate facilities that can demand large amounts of energy and material used in all phases of the life cycle.

- End-of-life management, covering the various EOL stages (and their accompanying legislation) and providing support in creating a framework for environmentally sound management of EOL ICT equipment.

- General specifications and key performance indicators, with a focus on the matching environmental Key Performance Indicators (KPIs) to an organization's specific business strategy targets, and the construction of standardized processes to make sure the KPI data is as useful as possible to management.
- Assessment framework for environmental impacts explores how the various standards and guidelines can be mapped so that an organization can create a sustainability framework that is relevant to their own business objectives and desired sustainability performance (Toolkit on Environmental Sustainability for the ICT Sector, 2012).

To achieve best practices, continuous improvement uses elements from the International Organization for Standardization (ISO), risk management frameworks, and other means consist of incremental initiatives and innovations. All expectations and objectives should be deployed to stakeholders clearly, promoting and supporting greater awareness, accountability, and transparency. TL 9000 is a globally recognized quality standard and is designed to improve telecommunications or ICT products and services. TL 9000 includes all of the requirements of the ISO 9001 quality standard, plus additional requirements that focus on customer needs, strategic planning, and accountability. ICT organizations should align their sustainability initiatives to TL 9000 certification, which in essence builds on a set of standardized metrics for performance, reliability, and delivery that provide industry-benchmarking capabilities (TL 9000 – Quality Certifications, 2010).

The TL 9000 standard is the telecom industry's unique extension to ISO 9001:2015 and includes supplemental requirements in the following areas:

- Performance measurements based on reliability of product;
- Software development and life-cycle management;
- Requirements for specialized service functions such as installation and engineering;
- Requirements to address communications between telecom network operators and suppliers;
- Reporting of quality measurement data to a central repository.

As a result, the TL 9000 is a two-part quality system with significant management *and* measurement components. In particular, TL 9000-certified organizations are required to comply with:

- All requirements clauses of the International Standards ISO 9001:2015;
- Telecom-specific requirements that apply to all registrations;

- Telecom-specific requirements that apply to hardware, software and/or service registrations;
- Telecom industry measurements that apply in all product categories;
- Telecom industry measurements that apply in certain product categories specific to hardware, software, and/or services (TL 9000 Quality Management System, 2017).

ICT's development of industry standards tailored to their sustainability operations improves reputation by scoring a company's assessment of credible international standards and evidence of performance against industry benchmarks. The particular efficiencies and benefits can be described as follow:

- **Alleviates supply chain risk** by measuring the percentage of the organization's suppliers engaged in sustainability and determines what minimal compliance looks like versus having a competitive advantage.
- **Reduces costs and increases profitability** through reduced resource consumption, increased worker productivity, reduced quality defect levels, and reduced product cost.
- **Increases sales** by enabling product and service innovation through the application of sustainability tools and concepts such as eco-design, circular economy, resource efficiency, and end-to-end supply chain optimization.
- **Assures industry best practices** using case studies and a consultative approach to standards, providing recommendations for improvement.
- **Aligns employees, shareholders, and management** with corporate initiatives by opening a two-way dialog between your organization and those affected by its activities (Quest Sustainability Assessor, 2017).

Intertwined between Biotechnology and Sustainable Development

The accelerated rate of the information being generated is unprecedented with the parallel accessibility of information. The aforementioned information accessibility, useful to the role of science and technology in economic development, is valuable as technology networks are transforming the traditional map of development, unifying countries, regions, populations, and creating a "Knowledge Society" with universal access to all knowledge creation. Within the sustainable development context, technological advances are a powerful tool for human development and poverty reduction as well as building sustainability in every aspect of our modern society. The hopes are that new technologies, such as ICT and biotechnology, will lead to healthier lives, greater social freedoms, improved knowledge, and more productive livelihoods (Science and Technology for Sustainable Development, 2003). The role of science and technology (S&T) for sustainable growth, poverty alleviation, and development is critical. Advances in scientific and technological knowledge made possible the significant reductions of poverty and improvements in the quality of life in the last century and more prominently in this century in both developed and developing countries. The scope for developing new emerging technologies and incorporating science can significantly influence the global discourse on social and economic development (Monyooe and Ledwaba, 2004).

Biotechnology can be summarized as technology based on biology. Biotechnology harnesses cellular and bimolecular processes with the aim of developing technologies and products that supports improving the well-being of people and promotes a healthy planet. Biological processes have been used for more than 6,000 years with many utilities such as making useful food products, preserving dairy products, etc. Bioprocessing techniques such as fermentation, baking, and tanning have been used throughout much of human history. In recent history, we have witnessed major advances made possible by techniques such as genetic engineering and the development of the biotechnology industry. Fermentation is arguably the earliest example of biotechnology and consists of the metabolic process by which microbes produce energy in the absence of oxygen and other terminal electron acceptors in the electron transport chain such as fumarate or nitrate. In ancient times, it was considered as a way to both preserve food and to retain nutritional value (Committee on Industrialization of Biology: A Roadmap to Accelerate the Advanced Manufacturing of Chemicals, 2015). Economic growth – as measured by gross domestic product (GDP) – is a key determinant in the growth of energy demand. Although consumption of non-fossil fuels is expected to grow faster than fossil fuels, fossil fuels will still account for 77% of energy use. Natural gas is the fastest-growing fossil fuel in the projections. Global natural gas consumption increases by 1.4%/year. Abundant natural gas resources and rising production are factors that place natural gas in a competitive position in the global markets. Liquid fuels – mostly petroleum-based – remain the largest source of world energy consumption (International Energy Outlook, 2017). Consequently, fossil fuels are still a major source of energy; although renewable energy usage is growing exponentially and plays an important role in reducing greenhouse gas emissions. When renewable energy sources are used, the demand for fossil fuels is reduced.

As shown in Figure 3.6, the world energy consumption increases from 575 quadrillion British thermal units (Btu) in 2015 to 663 quadrillion Btu by 2030 and then to 736 quadrillion Btu by 2040. Most of the increase in energy demand is expected to come from non-OECD countries, where strong economic growth, increased access to marketed energy, and quickly growing populations lead to a rising demand for energy. Energy consumption in non-OECD countries increased by 41% between 2015 and 2040, in contrast to a 9% increase in OECD countries (International Energy Outlook, 2017).

The following example represents some of the possible biotechnological paths relevant to sustainable development and closely related to finding alternative fuels:

Energy generation – Increased productivity of ethanol through genetic improvement of yeast. Ethanol production by fermentation is, by definition, a biotechnological process, as the agent responsible for transforming sugar into alcohol – the yeast *Saccharomyces cerevisiae* – is a living organism. The production of food and beverages among other products using yeast has been employed for thousands of years and implied an important economic relevance as the most widely used microorganism for the industrial production of ethanol, because of its high selectivity in ethanol production, high growth and fermentation rate, high ethanol yield, high tolerance to glucose, ethanol, osmotic pressure and stressful conditions, low optimum fermentation pH, and high optimum fermentation temperature (Amorim and Leao, 2005).

There are important factors to be considered affecting carbon dioxide emissions. CO_2 emissions are affected by several variables, particularly conversion efficiency and crop yield. Evaluating different scenarios and suppositions for such values will result in different values of CO_2 emissions. Ethanol derived from sugarcane, for example, places the agricultural inputs to be accountable for the larger amount of emissions; for ethanol derived

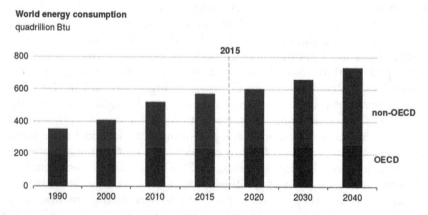

FIGURE 3.6
World energy consumption. (From U.S. Energy Information Administration, 2017.)

from corn, most emissions result from conversion. While using ethanol fuel has some environmental benefits, there are also detrimental environmental impacts, such as water use. A large quantity of water is used to clean sugarcane, because of the large amounts of soil attached to its stalks. Moreover, preharvest burning of sugarcane is related to increased levels of carbon monoxide and ozone in the agricultural region and cities where sugarcane is produced (Dias de Oliveira et al., 2005). These processes can present air quality problems that will directly affect the human population. Urban and industrial waste is a complex issue that affects communities and people's health. Urban populations had to deal with the waste dumped upstream; rural populations had to deal with the influx of urban waste. The prevention of environmental pollution needs to be tackled in order to improve the conditions where livestock manure is managed and prevent nitrates and other dangerous minerals from contaminating the soil, in addition to preventing air and water pollution. The production of biopolymers from renewable resources can help solve the issue of urban and industrial waste, by replacing plastics of petrochemical origin with plastic produced by microorganisms, since biopolymers are bio-compatible and fully biodegradable materials. Choi and Lee (1999) and Schenberg (2010) stated that the estimate for plastic waste dumped in landfills increases the total weight of the landfill waste by 404% and the problem is worsened by the fact that the plastic materials currently produced are slow to decompose, remaining in the environment for several hundred years. Nevertheless, the price of biopolymers is not yet able to compete with conventional plastics. It is needed to optimize the microbial strains, as well as the extraction and recovery processes, but particularly to reduce the costs of raw materials to make it viably competitive in the international markets with conventional plastics (Choi and Lee, 1999; Schenberg, 2010).

Smart Components and Smart Systems Integration

Electronic and photonic components, integrated micro/nanosystems, multicore computing systems, embedded systems with monitoring and controls are all cooperating complex systems that support the search for clean and renewable alternative energy resources as one of the most urgent challenges to the sustainable development of human civilization. The deep miniaturization, energy efficient, performance increase, and manufacturability of nano-electronic devices, using alternative solutions to the traditional miniaturization path for information and communication systems and other applications, are fundamental applications of the integration of new functionalities for the next generation of application-specific components and smart systems. Therefore, ICTs, as well as the convergence of micro-electronics, nanomaterials, biochemistry, biosensors, and measurement technologies, have the capacity of utilization for a wide field of applications such as biosensors which have acquired paramount importance in the field of drug discovery, biomedicine, food safety standards, defense, security, and environmental monitoring. This has led to the invention of precise and powerful analytical tools using biological sensing elements (Vigneshvar et al., 2016). The fields of medicine, biology, communications, lighting, sensing and measurement, and manufacturing can largely benefit from further development of core and disruptive photonic technologies, such as lasers, waveguides, photodetectors, amplifiers, LEDs, and optical fibers, which are now revolutionizing the ICTs. Furthermore, the development of advanced, low-temperature processing, and potentially printable devices and systems on large areas and/or flexible substrates, such as light emitting and sensing devices, photovoltaics, displays, printed electronics for smart tags, or wearable smart textiles. It is important to highlight the high-performance flexible devices like thin-film transistors, thin-film batteries, solar cells, and antennas that offer system-level solutions for low-cost mass production and energy-efficiency demands of future smart buildings; these devices can provide environmental benefits, minimizing health risks and reducing energy consumption and subsequently carbon emissions (Srivatsan and Sudarshan, 2016).

Micro/Nanotechnology-Enabled Technologies for Energy Harvesting

Nanotechnologies are technologies that contain at least 50% natural or manufactured particles in the size range from 1 to 100 nm (up to a billionth of a meter) (Parisi et al., 2014). Nanotechnology is helping to substantially improve, even revolutionize, many technology and industry sectors such as homeland security, medicine, transportation, energy, food safety, and environmental science. Nanotechnology has unique physical properties because of their large surface area and this dominates the contributions made by the small bulk of the material. Moreover, thanks to its potential for its unique properties (e.g. high volume to surface ratio and high solubility), nanoparticles show an increased reactivity and efficiency. This is however not a universal definition of nanotechnology, as definitions differ regarding size and properties of nanomaterials (Bakker et al., 2017).

Micro/Nanosystems and Renewal Energy

One of the most critical challenges of the twenty-first century to achieving sustainable development of human civilization is the utilization of clean and renewable alternative energy resources to cope with the growing threat of severe climatic events, pollution, and energy crisis (Science, 2007). Petroleum, coal, hydraulic power, natural gas, wind power, and nuclear plants are the main drivers of energy sources in our world today. However, efforts to the research and development of the exploration of alternative sustainable energy resources, such as solar energy, geothermal power, biomass/biofuel, and hydrogen energy are not fully developed and utilized for the large-scale supply of power. The potential for energy utilization from these sources is still mainly used for small-scale powering applications (Wang and Wu, 2012).

Harnessing renewable energy is an appropriate first consideration in sustainable development because there is no depletion of mineral resources and no direct air or water pollution. There is a variety of sources available for energy scavenging by micro/nanosystems (MNSs) from the ambient environment. Renewal energy technologies provide the tools to accomplish the SDGs to mitigate environmental impacts from energy systems. Natural forms of energy, including, but not limited to, energy in natural forms, such as wind, water flow, ocean waves, and solar power; mechanical energy, such as vibrations from machines, engines, and infrastructures; thermal energy, such as waste heat from heaters and joule heating of electronic devices; light energy from both domestic/city lighting and outdoor sunlight; and electromagnetic energy from inductor coils/transformers as well as from mobile electronic devices, all need to be fully developed and become crucial components of any transformation strategy (Wang and Wu, 2012). Solar energy is facing many challenges such as grid infrastructure not well-equipped to handle renewable energy, and inefficiency because the majority of energy from the sun's rays isn't even captured by solar systems. Nanotechnology can help overcome current performance barriers and substantially improve the collection and conversion of solar energy. Nanoparticles and nanostructures enhance the absorption of light, increase the conversion of light to electricity, and provide better thermal storage and transport.

In summary, the goals of nanotechnologies can be to:

1. Improve photovoltaic solar electricity generation;
2. Improve solar thermal energy generation and conversion;
3. Improve solar-to-fuel conversions (A Progress Review of the Nanotechnology for Solar Energy Collection and Conversion (Solar) NSI, 2015) (Figure 3.7).

The above figure is a nanotechnology solar cell which absorbs both sunlight and indoor light and converts it into electricity. The basic concept is that Plastic, a thin, flexible, solar panel, is made using nanoscale titanium particles coated in photovoltaic dyes, which generate electricity when they absorb light (Bharathidasan and Muhibullah, 2006).

Micro-Nano Bio Systems

The Micro-Nano Bio Systems (MNBS) concept sees the technological integration of its utilities that can range over a wide spectrum of applications from agricultural use – biosensors in food processing for example – with the aim to apply the state-of-the-art micro/nanotechnology to solve the practical

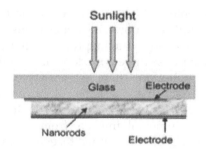

FIGURE 3.7
Diagram of a nano solar cell. (From Bharathidasan and Muhibullah, 2006.)

challenges existing in the science and engineering fields. It is necessary to tailor research and development (R&D) nanotechnology in food, health care and agriculture, green energy, and green technology with IT and communications in the same fields. The important point is to create the MNBS concept that is inclusive, that encourages integration, has applications in a wide range of markets, and avoids intellectual, market, and commercial fragmentation. The highly cross disciplinary nature of MNBS opens a wide range of new possibilities for intellectual advance and competitive commercial development. Micro-nano bio systems are a platform that is developing strongly, particularly in developed countries to address problems in a whole range of markets and to facilitate integrating MNBS into communities to deliver broader societal benefits.

The development and system integration across technologies like micro-nano electronics, nanotechnology, biotechnology, biomaterials, and ICT to achieve innovation in several sectors and applications (e.g., health care, food and agriculture, and environment and safety) is relevant to a sustainable society with a level of consumption that should reflect environmental and resource balance. As sustainable development should assure its citizens equality, freedom, and a healthy standard of living, the use of new technologies and in particular MNBS addressing societal challenges is vital to accomplishing the SDGs. Some of the benefits can be aligned as follows:

1. *Agriculture*: to determine the quality of the soil, pesticide and herbicide load, antibiotic and growth hormone levels.

2. *Food processing activity*: to identify and quantify microbes and metabolic toxins, poisonous heavy elements, and compounds that need to be minimized or eliminated to avoid the chronic poisoning of food products.

3. *Health care*: to support the assessment of patient response to drugs, the identification of bacterium/virus and determination of pathogen load, deliver subject monitoring at the point of care (POC), and provide wider support to public health with accurate, timely data.

4. *Environment*: to identify the occurrence of pollutants at the source and aid the separation and removal of undesirable materials from our living environment (Translating Technologies into Competitive, Validated & Manufacturable Products to Impact Quality of Life, 2016).

Sustainable development and technology advancements face many challenges and to overcome the long-term risks of its use and applications such as nanoparticles, which are nearly impossible to measure in our air, may explain why pollution is linked to heart attacks and stroke. The challenges adopting international regulations and standardized risk assessment tools, assessing its cost-efficiency, and the inclusive availability of nanotechnology applications to both developed and developing countries, MNBS solutions must demonstrate dependability in real life environments and cost-effectiveness in addition to minimizing health risks.

It is imperative that new technology devices and products are affordable, smart and autonomous, small, intuitive to use, and connected, delivering for example, fast and low-cost diagnosis, detection of hazards for health, parameters for enhanced food safety, and improved process control.

The agricultural sector is still comparably marginal and has not yet made it to the market to any larger extent compared with other sectors of nano-technology application. Mukhopadhyay (2014) stated that while the seeds of research in nanotechnology started growing for industrial applications nearly half a century ago, the momentum for use of nanotechnology in agriculture came only recently with the reports published by Roco (1999), United States Department of Agriculture (2002), the Nanoforum (2006), and Kuzma and VerHage (2006), along with similar publications (Mukhopadhyay, 2014) (Figure 3.8).

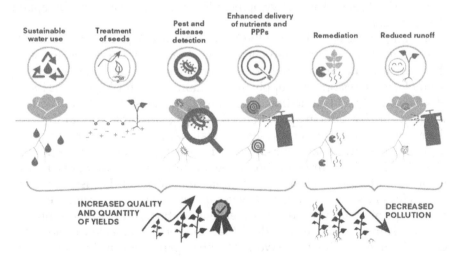

FIGURE 3.8
Benefits of nanotechnology in agriculture. (From Mukhopadhyay, 2014.)

The figure above shows the benefits of nanotechnology in agriculture. Agricultural nanotechnology is a tool that can provide greater dividends for poor nations because it is powerful in ameliorating problems related to poor input-use efficiency, water scarcity, poor sanitary conditions, and other similar problems experienced by poor nations. However, poor nations can harvest the fruits of nanotechnology if it is realized that future cost of importing farm- technology could be higher than that of developing it indigenously in a sustained manner.

Nanosystems for Water Quality Monitoring and Purification

Water-stressed countries are regions with fewer than 1,700 m^3 of water per person per year. The implications are profound, affecting global and local economic development, straining the environment, and drastically limiting food availability. Populations living in water-stressed regions suffer from multiple risks related to their well-being, in particular, health. The 2009 World Water Development report revealed that nearly half of the global population will be living in regions of high-water stress by 2030 (World Water Assessment Programme, 2009).

Minimizing health risks is closely related to the improvement of sanitation services. Therefore, the improvement of wastewater treatment is a first step to combat health hazards. In that context, the United Nations World Water Development Report (2017) published some statistics highlighting the vital importance of improving the management of wastewater for our common future. Some 2.1 billion people have gained access to improved sanitation facilities since 1990. However, 2.4 billion still do not have access to improved sanitation and nearly 1 billion people worldwide still practice open defecation (UN World Water Development Report: Wastewater, the Untapped Resource, 2017).

Building on the experience of the MDGs, the 2030 Agenda for Sustainable Development has a more comprehensive goal for water. It is a must to assess the present strategies to cope with water resources and utilize emerging technologies to restore and supply the communities, going beyond the issues of water supply and sanitation. SDG Target 6.3 states: By 2030, to improve water quality by reducing pollution, eliminating dumping, and minimizing the release of hazardous chemicals and materials, halving the proportion of untreated wastewater and substantially increasing recycling and safe reuse globally. There is an urgent action to be taken to address wastewater treatment and ICT as an enabler to smart water management. The issues related to an efficient system to improve water quality reveals an urgent need for technological upgrades and safe water reuse options, which are essential to accomplishing the entire Agenda (Water and the Sustainable Development Goals, 2015). According to the World Health Organization, in 2015, 71% of the global population (5.2 billion people) used a safely managed drinking-water service – that is, one located on premises, available when needed, and free from contamination. Some of the statistics of the report are:

- 89% of the global population (6.5 billion people) used at least a basic service. A basic service is an improved drinking-water source within a round trip of 30 minutes to collect water.
- 844 million people lack even a basic drinking-water service, including 159 million people who are dependent on surface water.
- Globally, at least 2 billion people use a drinking-water source contaminated with feces.
- Contaminated water can transmit diseases such as diarrhea, cholera, dysentery, typhoid, and polio. Contaminated drinking water is estimated to cause 502,000 diarrheal deaths each year.
- By 2025, half of the world's population will be living in water-stressed areas.
- In low- and middle-income countries, 38% of health-care facilities lack an improved water source, 19% do not have improved sanitation, and 35% lack water and soap for hand-washing (World Health Organization, 2018).

The strategy needs to be prioritized around developing sensors for monitoring water and wastewater quality with nanotechnology, offering the ability to control matter at the nanoscale and create materials that have specific properties with a specific function. Hence, new biological and electronic techniques and new long-life low power systems are key factors to improve monitoring and measurement technology in pipe sensors. Nanosensors for monitoring water quantity and quality in public water systems could lead to completely decentralized control and optimize the process. Increase water availability using nanotechnology. In essence, the three main points of the development of technological solutions that can relieve current stresses on the water supply and deliver methods to sustainably utilize water resources in the future are:

- Increase water availability using nanotechnology;
- Improve the efficiency of water delivery and use of nanotechnology;
- Enable next generation water monitoring systems with nanotechnology (Water Sustainability Through Nanotechnology: Nanoscale Solutions for a Global-Scale Challenge, 2016).

Nanomaterials have unique size-dependent properties. Specifically, nanomaterials deal with sizes of 100 nm or smaller in at least one dimension. Therefore, the material properties of nanostructures are different from the bulk due to the high surface area over volume ratio and the possible appearance of quantum effects at the nanoscale. In this case, these properties are ideal for treating nontraditional water sources. They have the potential to enable the development of novel nanotechnology-based solutions for more efficient utilization of drinking water, nontraditional water sources, and wastewater treatment processes. Water treatment processes are essential due to the population increase that leads to contamination of the water systems; wastes from human activities contaminate most of the water supplies, among other issues. Water treatment plays an important role to properly treat a contaminated source of water supply in order to protect the health of populations. In this case, many nanotechnology-enabled approaches can be useful in the water treatment process. For example, membranes are thin sheets of material that are able to separate contaminants based on properties such as size or charge (National Academy of Sciences. 2007). They can be designed with nanoscale pores that remove specific pollutants while allowing water molecules and important nutrients to pass through, and the antimicrobial properties of silver nanoparticles can be utilized for point-of-use water disinfection. Furthermore, the broad application space for nanotechnology-enabled water treatment and the adaptation of highly advanced nanotechnology to traditional process engineering presents opportunities for development of advanced water and wastewater technology processes to obtain the required quality control (Table 3.8).

Nanomaterials have unique size-dependent properties related to their high specific surface area (fast dissolution, high reactivity, strong sorption) and discontinuous properties (such as superparamagnetism, localized surface plasmon resonance [SPR], and quantum confinement effect). The aforementioned properties facilitate novel high-tech materials for more efficient water and wastewater treatment processes, namely membranes, adsorption

TABLE 3.8

Overview of Types of Nanomaterials Applied for Water and Wastewater Technologies

Nanomaterial	Properties	Applications
Nanoadsorbents	+ high specific surface, higher adsorption rates, small footprint – high production costs	Point-of-use, removal of organics, heavy metals, bacteria
Nanometals and nanometal oxides	+ short intraparticle diffusion distance compressible, abrasion-resistant, magnetic + photocatalytic (WO_3, TiO_2) – less reusable	Removal of heavy metals (arsenic) and radionuclides, media filters, slurry reactors, powders, pellets
Membranes and membrane processes	+ reliable, largely automated process – relative high energy demand	All fields of water and wastewater treatment processes

Source: Gehrke et al. (2015).

materials, nanocatalysts, functionalized surfaces, coatings, and reagents. The above table shows the most promising materials and applications of nano-materials applied to water and wastewater technologies (Gehrke et al., 2015).

In summary, nanotechnology can be an important tool to attain sustainable development addressing global sustainability in energy, water, food, shelter, transportation, and health care. The potential social benefits are greater than its challenges, taking into consideration practical actions to reduce its impact on the environment and climate. The potential for nanotechnology to address many systemic issues related to global sustainability should be weighed against uncertainties related to the environmental and health effects of nanomaterials (Lowry et al., 2010).

Nanotechnology and Smart Cities

Smart cities and nanotechnologies are inexorability linked in the future of humankind as nanotechnology is a cross-cutting issue, and the research and developing of new technologies will intertwine the two and evolve sustainability initiatives and efficiency above other priorities. The use of smart technologies combined with city planning led to the developing of smart cities with the aim of achieving modern urban systems with efficient tools to cope with growing needs from increasing population sizes. In essence, integrating ICT and IoT solutions to assess the risks and opportunities to minimize disruptions managing the city's assets can create a smart city that is environmentally friendly and more economically viable, thereby addressing some of the challenging problems faced in cities (Vandenbroucke, 2017). Recent advances in nanotechnologies have enabled a new family of sensors, termed self-sensing materials, which would provide smart cities with the means to monitor the structural health of civil infrastructures. There are several studies on the use of microparticles, nanoparticles, and composites, with a focus on the development of smart materials with innovative functionalities.

Prospective Applications in Smart Cities

Carbon nanotubes (CNTs) cementitious sensors represent a novelty in the field of structural engineering and Structural Health Monitoring (SHM). There are potential applications that are appropriate for fabricating smart self-sensing sensors, with the benefit of improving the city's operations and management. To accomplish this, these nanocomposites distribute large arrays of distributed sensors at low cost (D'Alessandro et al., 2015). Environmental monitoring is highly significant in addressing toxic contaminants and pathogens that are released through air, soil, and water. Sensors are devices that detect physical, chemical, and biological signals and provide a way for those signals to be measured and recorded (Hellman H. *Beyond Your Senses*. New York: Lodestar Books; 1997). These devices can monitor temperature, pressure, vibration, sound level, light intensity, load or weight, flow rate of gases and liquids, amplitude of magnetic and electronic fields, and concentrations of many substances in gaseous, liquid, or solid states. Furthermore, they have capabilities to keep the patients healthy for as long as possible, and they can enable bedside and remote monitoring of vital signs and other health factors. In more advance stages, robotic networks of platforms carrying physical, chemical, and biological sensors can monitor basic metabolic processes to assess ocean health and help track the ocean carbon, water's temperature, salinity, and acidity (Johnson, 2017).

The health of populations is being improved by new technologies with the help of advances in microchip technologies and molecular chemistry creating a new breed of microdevices for medical diagnostics. The development of simple and low-cost chemical sensors, which changes chemical information from the environment, is advancing rapidly and many types of chemical sensors have been developed in a variety of research areas such as air pollution in cities and people's exposure to environmental factors ranging from noise to particulate matter and gases, temperatures, humidity, and more (Tang, 2017). However, current sensors are not capable of measuring ultra-fine particles that pose an even greater risk to human health (HEI, 2013). The main technological challenge regarding the use of sensors for air pollution monitoring is to improve sensor performance – their sensitivity, stability, and longevity of operation before replacement (Serena, 2015). The implementation of guidelines and standards to enable international expansion, increase the efficiency and support development of new products and markets will be crucial for new technological advances.

List of Terms

Atalla Key Block (AKB): An advanced key management system.

Biodegradable: Able to decay naturally and without harming the environment.

Biodiversity: Biological diversity in an environment as indicated by numbers of different species of plants and animals.

Biotechnology: The use of biological processes, organisms, or systems to manufacture products intended to improve the quality of human life. The earliest biotechnologists were farmers who developed improved species of plants and animals by cross pollination or cross breeding.

British thermal unit (Btu): A measure of the quantity of heat, defined as approximately equal to 1,055 joules, or 252 gram calories.

Business continuity management (BCM): A framework for identifying an organization's risk of exposure to internal and external threats.

Carbon dioxide equivalent or CO₂e: A term for describing different greenhouse gases in a common unit. For any quantity and type of greenhouse gas, CO_2e signifies the amount of CO_2 which would have the equivalent global warming impact.

Carbon footprint: The total amount of greenhouse gases produced to directly and indirectly support human activities, usually expressed in equivalent tons of carbon dioxide (CO_2).

Carbon nanotubes (CNTs): Allotropes of carbon with a cylindrical nanostructure.

Cementitious sensors: Cement-based sensors that can be embedded into structural elements and therefore provides a distributed sensing system.

Cybersecurity: A set of techniques used to protect the integrity of networks, programs, and data from attack, damage, or unauthorized access.

Database management system (DBMS): System software for creating and managing databases. The DBMS provides users and programmers with a systematic way to create, retrieve, update and manage data.

Data Basin: A web-based platform that integrates science, mapping, and people. Data Basin supports users with a wide range of technical capabilities.

Decision support system (DSS): A computerized information system used to support decision-making in an organization or a business. A DSS lets users sift through and analyze massive reams of data and compile information that can be used to solve problems and make better decisions.

Dialog generation and management system (DGMS): A software package whose function in the dialog subsystem of a DSS development environment is similar to that of a DBMS in a database.

Digital exclusion: A discrete sector of the population suffers significant and possibility indefinite lags in its adoption of ICT through circumstances beyond its immediate control.

Disaster risk management: The application of disaster risk reduction policies and strategies to prevent new disaster risk, reduce existing disaster risk and manage residual risk, contributing to the strengthening of resilience and reduction of disaster losses.

Ecoinformatics: Also known as ecological informatics is the science of information (Informatics) in ecology and environmental science. It integrates environmental and information sciences to define entities and natural processes with language common to both humans and computers.

E-commerce (electronic commerce or EC): The buying and selling of goods and services, or the transmitting of funds or data, over an electronic network, primarily the Internet. These business transactions occur either as business-to-business, business-to-consumer, consumer-to-consumer or consumer-to-business.

Electronic waste or e-waste: Describes discarded electrical or electronic devices. Used electronics which are destined for reuse, resale, salvage, recycling, or disposal are also considered e-waste.

Energy Intensity: The portion of the total energy supply required to produce a material. For example, the manufacture of 1.5 gigatons of steel would have gobbled up one-fifth of the world's total primary energy supply (TPES) in 1900. In 2010 it used only about one-fifteenth (Graph 3.2).

Ferrimagnetic material: One that has populations of atoms with opposing magnetic moments, as in antiferromagnetism; however, in ferrimagnetic materials, the opposing moments are unequal and a spontaneous magnetization remains.

Ferromagnetic: The property of being strongly attracted to either pole of a magnet.

Fumarate: A salt or ester of fumaric acid.

Gigaton (Gt): A unit of explosive force equal to one billion (10^9) tons of trinitrotoluene (TNT).

Geographic information system (GIS): A system designed to capture, store, manipulate, analyze, manage, and present spatial or geographic data.

Health IT (health information technology): The area of IT involving the design, development, creation, use, and maintenance of information systems for the health-care industry.

Industrial ecology (IE): The study of material and energy flows through industrial systems. Industrial ecology has been defined as a systems-based, multidisciplinary discourse that seeks to understand emergent behavior of complex integrated human/natural systems.

Information system (IS): An organized system for the collection, organization, storage, and communication of information. More specifically, it is the study of complementary networks that people and organizations use to collect, filters, and process, create and distribute data.

Intermodality: Involving two or more different modes of transportation in conveying goods.

International Organization for Standardization (ISO): An international standard-setting body composed of representatives from various national standards organizations. Founded on 23 February 1947, the organization promotes worldwide proprietary, industrial and commercial standards.

International Organization for Standardization 22301 (ISO 22301): A proposed standard that specifies security requirements for disaster recovery preparedness and business continuity management systems (BCMS).

Internet of Things (IoT): The network of physical devices, vehicles, home appliances and other items embedded with electronics, software, sensors, actuators, and connectivity which enable these objects to connect and exchange data.

Knowledge base management system (KBMS): A computer application for managing (creating, enhancing, and maintaining) the AKB, just as a DBMS is a computer application for managing a database.

Landscape ecology: The study of the pattern and interaction between eco-systems within a region of interest, and the way the interactions affect ecological processes, especially the unique effects of spatial heterogeneity on these interactions.

Low-carbon economy (LCE): Also known as low-fossil-fuel economy (LFFE), or decarbonized economy is an economy based on low-carbon power sources that therefore have a minimal output of greenhouse gas (GHG) emissions into the biosphere, but specifically refers to the greenhouse gas carbon dioxide.

Micro-Nano Bio Systems (MNBS): A cluster of projects targeting systems and applications that have, or interact with, biological components.

Model base management system (MBMS): The activity of managing and making informed decision regarding the future direction of a business, process, or system(s) based on information gleaned and understood from models that document the current state.

Molecular chemistry: A creative science, where chemists synthesize molecules with new biological or physical

Multicore/multicore computing systems: An architecture in which a single physical processor incorporates the core logic of more than one processor. A single integrated circuit is used to package or hold these processors.

Nanoparticles: Particles between 1 and 100 nanometers (nm) in size with a surrounding interfacial layer.

Nanostructure: A structure of intermediate size between microscopic and molecular structures.

Nanosystem: Any physical system that is engineered at the nano scale.

Network society: The expression coined in 1991 related to the social, political, economic and cultural changes caused by the spread of networked, digital information and communications technologies (ICTs).

Nitrate: A salt or ester of nitric acid.

NOx: The gases nitric oxide and nitrogen dioxide, that are produced when fuel is burned and that are harmful to the environment.

PDCA (plan-do-check-act, sometimes seen as plan-do-check-adjust): A repetitive four-stage model for continuous improvement (CI) in business process management. The PDCA model is also known as the Deming circle/cycle/wheel, Shewhart cycle, control circle/cycle, or plan-do-study-act (PDSA).

Photonics: The physical science of light (photon) generation, detection, and manipulation through emission, transmission, modulation, signal processing, switching, amplification, and detection/sensing.

Photovoltaic: Relating to the production of electric current at the junction of two substances exposed to light.

Risk management: The process of identifying, assessing and controlling threats to an organization's capital and earnings. These threats, or *risks,* could stem from a wide variety of sources, including financial uncertainty, legal liabilities, strategic *management* errors, accidents and natural disasters.

Satellite radar altimeters measure: An artificial object which has been intentionally placed into orbit to measure the ocean surface height (sea level) by measuring the time it takes a radar pulse to make a round-trip from the satellite to the sea surface and back.

Superparamagnetism: A form of magnetism which appears in small ferromagnetic or ferrimagnetic nanoparticles. In sufficiently small nanoparticles, magnetization can randomly flip direction under the influence of temperature.

Surface plasmon resonance (SPR): The resonant oscillation of conduction electrons at the interface between negative and positive permittivity material stimulated by incident light.

Tele-education: An education in which the students receive instruction over the Internet, from a video, etc., instead of going to school; it is also known as e-learning which comprises all forms of electronically supported learning and teaching systems.

Terrestrial ecosystem: A type of ecosystem found only on biomes, also known as beds. Six primary terrestrial ecosystems exist: tundra, taiga, temperate deciduous forest, tropical rain forest, grassland, and desert.

The Brundtland Commission: Formerly known as the World Commission on Environment and Development, with the mission to unite countries to pursue sustainable development together. It was officially dissolved in December 1987 after releasing Our Common Future, also known as the Brundtland Report, in October 1987; a document which coined and defined the meaning of the term "Sustainable Development."

Urbanization: Population shift from rural to urban areas; the gradual increase in the proportion of people living in urban areas, and the ways in which each society adapts to the change.

4

Health in the Digital World

Health-Care Transformation

Health Information Technology (HIT) and Digital Health are rapidly transforming health-care systems spurred by technology innovation, government initiatives, and growing challenges of the twenty-first century in improving quality, efficiency, and patient experiences. Furthermore, the transformation of the health-care system through new technological developments is moving into processes, practices, and relationships across the ecosystem, creating new opportunities for the sector and the potential to develop new strategies to cope with health risks.

mHealth: A New Path of Transformation for Health through Mobile Technologies

According to the International Telecommunication Union, which is a specialized agency of the United Nations (UN), 7 billion people (95% of the global population) live in an area that is covered by a mobile-cellular network. Mobile-broadband networks (3G or above) reach 84% of the global population, but only 67% of the rural population. Long-Term Evolution (LTE) networks have spread quickly over the last three years and reach almost 4 billion people today (53% of the global population), enhancing the quality of Internet use (ICT: Facts and Figures, 2016).

The World Health Organization second global survey on eHealth, *mHealth: New horizons for health through mobile technologies* (2011), published new highlights about mHealth that is defined by small-scale pilot projects addressing single issues on information sharing and access. This new path of health transformation has the potential to play a key role in health care with the support of public-private partnerships to implement large-scale and complex programs that will become more common as mHealth matures and the policies and strategies to support its benefits are in place. Governments and private institutions are expressing interest in mHealth as a complementary strategy for strengthening health systems. For example, these initiatives act as either stand-alone aids for Veterans with post-traumatic stress disorder (PTSD) or adjuncts to conventional psychotherapy approaches. These initiatives include maternal and child health, and evolve into a series of mHealth deployments worldwide that are providing early evidence of the potential for mobile and wireless technologies. These technologies can also aid in reducing the burden of the diseases linked to poverty including HIV/AIDS, malaria, and tuberculosis (TB) (Ben-Zeev, 2016) (Figure 4.1).

Many of these diseases, including HIV/AIDS, tuberculosis, and malaria are often considered "diseases of poverty"; therefore, mortality rates and burdens of disease are higher in less developed countries. In addition, malaria and dengue are water-related vector-borne diseases, and global evidence supports a direct and consistent association between access to clean water and poverty levels (UN-Habitat: The Challenge of Slums Global report on Human Settlements, 2006). Consequently, addressing health risks with mobile technologies helps to shorten the gap to environmental conditions and poverty, which affect the lives of the poor disproportionately, posing serious health risks and a restrained economy. Although the world is experiencing health advances in many areas, poor health continues to be a constraint on sustainable development efforts.

Fighting Chronic Disease through Mobile Coaching

Chronic diseases and conditions such as heart disease, stroke, cancer, type 2 diabetes, obesity, and arthritis are among the most common, costly, and preventable of all health problems and are also the leading causes of death

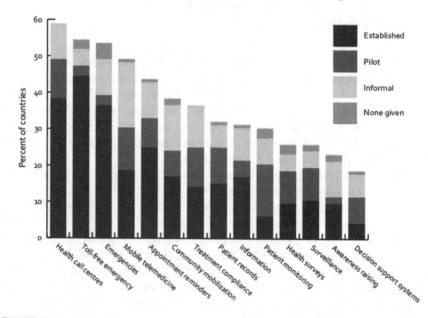

FIGURE 4.1
Adoption of mHealth initiatives and phases, globally (mHealth New Horizons for Health through Mobile Technologies, 2011).

and disability globally. As health care has become more complex, the burden on most health-care systems has become more acute. An example of the benefits in prevention and early intervention as a big step toward the ultimate goal of making populations healthier through better lifestyles and increased compliance with their suggested care regimens is the impact mobile phone-based coaching and online decision support can have on diabetes patients. Chronic disease management often requires a long-term care plan; hence, chronic disease management is critical to achieving improved health outcomes, quality of life, and cost-effective health care (Hamine et al., 2015). The potential benefit of mHealth health-care treatment and prevention to specially target chronic disease patients with customized sensors, devices, services, and tools to modify behavior in an engaging and sustainable way, is gaining more acceptance. These devices and services permit patients to monitor their health, access health information, and communicate with their health-care provider without requiring a wired connection to the Internet and facilitate the self-management and care coordination tool. The former is largely clinical; the latter is contingent on numerous and varying social determinants of health.

WellDoc, a health-care behavioral science and technology company, developed an application called BlueStar®, the first-of-its-kind Mobile Prescription Therapy cleared by the FDA and available only by prescription. It provides patients with real-time guidance to improve their diabetes self-management, as well as clinical decision support. The advantages of personalized health-care advice take away the health-care management from its current reactive and disease-centric system to one that is proactive and focused on wellness, disease prevention, and the precise treatment of disease. These tools help doctors optimize their diabetes treatment plan among other benefits such as obtaining health care without the supervision of a licensed health-care provider.

A 2009–2010 US trial of the WellDoc system sought to reduce blood glucose levels in 163 patients suffering from diabetes (Mobile Coaching and Online Decision Support for Diabetes, 2014; Quinn et al., 2017). The trial involved patients receiving a glucose meter and supplies. They also have access to the web-based portal and a mobile phone. Patients entered blood glucose levels and other self-care data into their phones – both "feature" as well as "smart" – and receive real-time responses from a "virtual patient coach." The assistance come through an expert mobile/web-client and cloud-based software system, providing management conditions, guidance in diets, physical exercise, and improvement in the general lifestyle. Some other benefits of the system include quality measurement and reporting, that produces more accurate and useful information for all health-care personnel, and evidence-based reports to monitor their conditions more accurately. Figures from the Centers for Disease Control and Prevention (2010) and the American Diabetes Association, (2009) estimates $218 billion is spent annually on diabetes in the United States. The potential for these mobile solutions could enable annual cost savings per patient of as much as US$10,000 in reduced health-care charges and increased worker productivity (Milliman, 2011; Roebuck et al.; Fraze et al. 2008; Rockville (MD): Agency for Health Care Policy and Research (US), Aug 2006–2010; Testa et al., 1998).

The results of the trial involving patients receiving a glucose meter and supplies have shown:

- A mean decline in A1c (glycated hemoglobin – the gold-standard measure for diabetes control) by 1.9% in the intervention group (against 0.7% in the usual care group) (Sacks, 2012).

- A clinically significant change in A1c was seen, regardless of whether patients began the trial with a high or low A1c. By comparison, the US Food and Drug Administration considers a new drug that is able to reduce A1c by 0.5% as clinically significant (Mobile health market, Report 2011–2016).

- WellDoc's application works on the vast majority of data-enabled mobile phones and can be integrated into the standard software and electronic health records used by doctors (Mobile health market, Report 2011–2016).

- The study also has shown a considerable application potential based on remote coaching for other chronic diseases such as diabetes, obesity, and hypertension. The aforementioned chronic diseases potentially have the highest utility for mobile management through smartphones and web-based solutions (Emerging mHealth: Paths for Growth, 2015).

Health Information Technology and Disease Prevention and Control

The rapid growth in health communication strategies and HIT to improve population health outcomes, health-care quality, and to achieve health equity is helping the health sector to cope with disease prevention and control, and disseminating information to educate patients and communities. However, the majority of US hospitals do not yet have HIT fully implemented to support effective communication between nurses, physicians, and the general population. Therefore, it is imperative to develop more tools and integrate a variety of electronic methods that are used to manage information about people's health and health care.

The Centers for Disease Control and Prevention (CDC) works to prevent chronic diseases and their risk factors through the following four domains:

- Epidemiology and surveillance refer to systems that are used to track chronic diseases and their risk factors.

- Environmental approaches refer to changes in policies and physical surroundings to make the healthy choice the easy choice.

- Health-care system interventions refer to improvements in care that allow doctors to diagnose chronic diseases earlier and to manage them better.
- Community programs linked to clinical services refers to those that help patients prevent and manage their chronic diseases with guidance from their doctor (The Four Domains of Chronic Disease Prevention, Centers for Disease Control and Prevention, 2015).

The use of ICT in epidemiology and surveillance is making a positive impact to support early detection, surveillance, and clinical management of diseases. The dengue virus project in Nicaragua – with the involvement of the Sustainable Science Institute (SSI) in collaboration with the MINSA Nicaragua, and supported by the Bill & Melinda Gates Foundation and the Fundación Carlos Slim para la Salud – involved the utilization of several ICT Tools:

- DENGUE – A L E R T: An automated early-alert dashboard for enhanced detection and response (by multiple types of users) to dengue outbreaks. This tool complements existing systems and incorporates information from a wide range of sources not currently incorporated into traditional surveillance systems in an efficient and direct way. Eventually, all priority diseases to be tracked at a national level will be incorporated into the ALERT system.
- DENGUE – L A B: An automated laboratory information management system to support sample management tracking and integration of results reporting for a network of national reference laboratories.
- DENGUE – S P E C I A L I S T: A mobile clinical decision support tool for doctors and nurses to facilitate hospitalized patient management with dengue. SPECIALIST is designed for use patient-side on tablets, cell phones, laptops, and/or ward-workstation laptops, depending on the hardware available (Sustainable Science Institute, 2014).

Another case study in the use of ICT tools is in India. The country has made considerable advancements in technology and keeps making progress in its commitment for establishing and operating a disease surveillance program responsive to the requirements of the International Health Regulations (IHR, 2005) by the World Health Organization (WHO). India is effectively using ICT for collection, storage, transmission, and management of data related to disease surveillance and effective response. Some of these advances are the establishment of terrestrial and/or satellite-based linkages within all states, districts, state-run medical colleges, infectious disease hospitals, and public health laboratories. The benefits of its usage translate into speedy data transfer, video-conferencing, training and e-learning for outbreaks, and program monitoring, as well as a media scanning and verification cell functions to receive reports of early warning signals. Moreover, an established call center offers a disease alert service that is open to the public seven days a week. The potential for ICT networks and efficacies was demonstrated in the 2009 H1N1 swine flu strain outbreak. The application of information and communication technologies (ICTs) was in place and part of its Integrated Disease Surveillance Project (IDSP). The utilization of ICTs was to further expand to hard-to-reach populations, to increase the involvement of the private sector, and to increase the use of other ways to share vital information and communication like e-mail and voicemail (Kant et al., 2010).

The Risks of Using Health Information Technology

It is understandable that more ICT will be deployed in the next few years than ever before, and will carry a variety of benefits, developing new paths for health-care treatment and management. Nevertheless, these developments do have risks to patients, leading some to call this a "dangerous decade" for HIT (Coiera et al., 2016). Some of the risks are poor communication between physicians and nurses, which is well known as one of the most common causes of adverse events for hospitalized patients and a major source for all sentinel events (Gawande et al., 2003; Leape and Berwick, 2005). The use of ICTs is increasingly changing the way physicians and nurses are communication and exchanging information. There is already evidence that communication technologies can contribute paradoxically to more issues and communication difficulties (Chiasson et al., 2007; Sutcliffe et al., 2004). It is crucial to understand the use of ICTs and its applications in health care; how it can be an instrument to eliminating barriers to quality and safety through increased awareness and its advantages to achieving the goals of better communication and safer care (Manojlovich et al., 2015).

Telemedicine, Telehealth, and Telecare

The incorporation of new technologies into the fields of health and social care are developing more than in any other field, and in the future it will continue to improve in impressive ways. While we can view these technological advances as promising and sometimes risky and debate the details of future trends in health care, we need to be clear about the drivers so we can align with them and efficiently work to guarantee the best benefits out of these developments in technology and the well-being of society as a whole. The main goal in eHealth is the efficacy, cost-effectiveness of the technology, management ratio, and delivery of vital information for the parties involved. Research supporting the usefulness of data collected by lifestyle monitoring systems is required to justify the associated intrusion, particularly in users with cognitive impairment. Studies thus far have focused on patient satisfaction and feasibility rather than the aforementioned benefits. There is an urgent need of research supporting the usefulness of data collected by lifestyle monitoring systems. Therefore, more development improvements of the health system can move further to justify the associated intrusion, particularly in users with cognitive impairment. Wootton et al. (2006) stated that this group may have the most to gain from devices designed to improve safety in the home; allowing them to live independently, but are at risk of losing their autonomy. Older people are likely to be disproportionately affected by technological change and geriatricians must be aware of the wide-ranging implications for their patients and practice. Hence, these tools, if developed properly according to the needs of patients and ability to deliver an efficient management service, can be a solution to problems of access to health-care services and as tools for the management of demand, especially for specialist services. Policy-makers must advocate for the increasing use of ICTs for improvements in the efficiency of health service provisions, which places emphasis on increasing responsibilities for patients to engage in self-care by mobilizing electronic resources (Kendall, 2009; Stowe and Harding, 2010).

Telemedicine Changing Health-Care Information Technology

The improvement and efficient performance of mobile technologies, more mature electronic health records (EHRs), and clinical decision support (CDS) systems are the most significant reasons for the growing interest of the health-care industry in telemedicine. The applications of telemedicine vary and can be deployed by a variety of medical providers and specialties. For example, some of the benefits are digital imaging and high-bandwidth communication to remotely view patient medical images – such as photos of skin lesions or CT scans – for diagnosis and treatment recommendations, which are under the dermatology and radiology medical specialties (Clinical Decision Support, 2017).

Another use of telemedicine is for those living in remote areas, or having experienced a disaster and not having access to specialists. Natural and man-made disasters are affecting populations around the world. The Centre for Research on the Epidemiology of Disasters (CRED) works on these four main areas:

- Civil strife and conflict epidemiology;
- Database and information support;
- Capacity building and training;
- Natural disasters and their impacts.

CRED report (2017) published the following data:

1. In 2016, 342 disasters triggered by natural hazards were registered.
2. Last year, the number of deaths caused by natural disasters (8,733) was the second lowest since 2006, largely below the 2006–2015 annual average (69,827).
3. Inversely, the number of people reported affected by natural disasters (564.4 million) was the highest since 2006, amounting to 1.5 times its annual average (224 million).
4. The estimates of natural disaster economic damages (US$154 billion) place in 2016 as the fifth costliest since 2006, 12% above the 2006–2015 annual average (Guha-Sapir et al., 2017).

Disasters continue to contribute to increased morbidity and mortality with significant economic impacts worldwide. Psychological, physical, and social sequelae persist years after the events (Nicogossian and Doarn, 2011). Many preexisting detrimental health and socio-economic conditions are exacerbated following disasters such as financial losses, misplacement, homelessness, mental health issues, anxiety, and in some cases PTSD. During a disaster, the communication network, roads, and access to emergency centers and hospitals may be overwhelmed or even disrupted, exacerbating the hazards and impeding the professional health-care team to reach out to the population affected by the impact; therefore, telemedicine can be a tool to use effectively and judiciously in the aftermath of disasters caused by both humans and naturally occurring events. The lack of effective communication after a disaster to collect progression and transmit important information is one of the biggest challenges facing the health-care team. After a disaster event an effective response is vital for survival and establishing rapid and reliable telecommunications systems specifically directed toward the "disaster medical field" is one of the most important steps to assure a prompt emergency response where mortality rates can be reduced if information exchange is dependable, fast, and accurate (Doarn et al., 2017).

Figure 4.2 describes the overall telemedicine system architecture. In each different application the telemedicine unit is located at the patient's site, whereas the base unit (or doctor's unit) is located at the place where the signals and images of the patient are sent and monitored. The telemedicine device is responsible to collect data (bio-signals and images) from the patient and automatically transmit them to the base unit. The base unit is comprised of a set of user-friendly software modules, which can receive data from the telemedicine device, transmit information back to it, and store important data in a local database. The system has several different applications (with small changes each time), according to the current health-care provision nature and needs. Before the system's technical implementation, an overview of the current trends and the needs of the aforementioned telemedicine applications

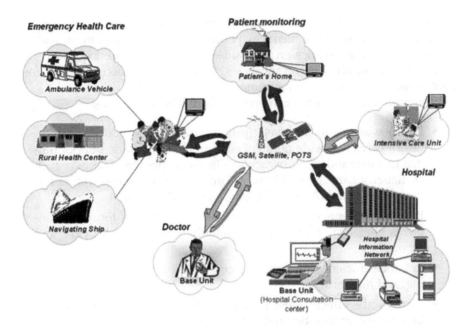

FIGURE 4.2
Telemedicine system architecture (Kyriacou et al., 2003).

were made, so that the different requirements are taken into account during design and development, thus ensuring maximum applicability and usability of the final system in distinct environments and situations.

As mentioned above, the system consists of two separate modules (Figure 4.1):

1. the unit located at the patient's site called "telemedicine unit," and
2. the unit located at doctor's site called "Base unit."

The doctor might be using the system either in an emergency case or when monitoring a patient from a remote place. The design and implementation of the system were based on a detailed user requirements analysis, as well as the corresponding system functional specifications (Kyriacou et al., 2003).

There are several telemedicine technology benefits aligned with sustainable development goals (SDGs) to ensure healthy lives and promote well-being for all ages. Therefore, telemedicine encompasses the electronic acquisition, processing, dissemination, storage, retrieval, and exchange of information. SDGs emphasize the promotion of a healthy society, which includes preventing disease, treating the sick, managing chronic illness, rehabilitating the disabled, and protecting public health and safety. Telemedicine strengthens the health-care system, creates partnerships, and improves the exchange of communication between different actors such as physicians, nurses, medical assistants, health emergency teams, and patients. Telemedicine offers an effective connectivity and exchange of information for providers to effectively connect with their patients by phone, e-mail, and webcam, providing quickness and convenience on both sides of the health-care experience, in addition to addressing important issues like inequities in access to care, cost containment, and quality enhancement.

In summary, telemedicine can provide the following benefits:

- Improve access to all levels (primary, secondary, and tertiary) of health care for a wide range of conditions—including, but not limited to, heart and cerebrovascular disease, endocrine disorders such as diabetes, cancer, psychiatric disorders, and trauma; as well as services such as radiology, pathology, and rehabilitation.
- Promote patient-centered care at lower cost and in local environments that also contributes to stabilizing local health care and economies.
- Enhance efficiency in clinical decision-making, prescription ordering, and mentoring.
- Increase the effectiveness of chronic disease management in both long-term care facilities and in the home.
- Promote individual adoption of healthy lifestyles and self-care (Rashid et al., 2009).

Telehealth, Population Aging, and Sustainable Development

Population aging is a phenomenon that results from declines in fertility as well as increases in longevity, two trends that are usually associated with social and economic development (Population Ageing and Sustainable Development, 2014).

The increasing prevalence of chronic disease among an aging population is presenting new challenges in the medical field and society as a whole. The development of new models of health-care delivery is foreseeable and the trend is likely to be focused in increasing prominence on patient self-monitoring, health-care delivery at patient homes, interdisciplinary treatment plans, a greater percentage of medical care delivered by non-physician health professionals, targeted health education materials, and greater involvement and training of informal caregivers. The information technologies (IT) infrastructure of health systems will need to adapt to this new reality and one of the possible solutions for this adaptation strategy is the use of telehealth as a possible approach to this problem. Telehealth is different from telemedicine because it refers to a broader scope of remote health-care services than telemedicine, and involves the remote exchange of data between a patient and health-care professional as part of the patient's diagnosis and health-care management (Beck et al., 2014). Telehealth can be used in reaching patients in their homes through remote monitoring where personal health and medical data is collected from a patient, monitoring blood pressure and blood glucose, and also being touted as a means to improve access to care, while reducing costs of transportation and increasing convenience to patients in obtaining care (Dixon et al., 2008). Favorable value propositions in the widespread use of telehealth involve a better understanding of patients health conditions by providing tools for self-monitoring; encourage better self-management of health problems, and alert professional support if devices signal a problem (Telehealth to Digital Medicine, 2014). In its many forms, telehealth offers conveniences including increased care accessibility and real-time synchro-nous audio-video encounters, and it presents the opportunity to reverse the long-standing standard of placing the burden on patients. Furthermore, it provides decreased transportation barriers and patient empowerment, bringing health care directly to the patient (Telehealth to Digital Medicine: How 21st Century Technology Can Benefit Patients, 2014). Steventon et al. (2014) published a comprehensive description on the use of telehealth, "Effect of Telehealth on Use of Secondary Care and Mortality: Findings from the Whole System Demonstrator Cluster Randomised Trial" is a home-based telehealth intervention report on the use of secondary health care and mor-tality and presented the following conclusions:

Objective: To assess the effect of home-based telehealth interventions on the use of secondary health care and mortality.

Design: Pragmatic, multisite, cluster randomized trial comparing telehealth with usual care, using data from routine administrative datasets. General practice was the unit of randomization. We allo-cated practices using a minimization algorithm, and did analyses by intention to treat.

Setting: 179 general practices in three areas in England.

Participants: 3,230 people with diabetes, chronic obstructive pulmo-nary disease, or heart failure recruited from practices between May 2008 and November 2009.

Interventions: Telehealth involved the remote exchange of data between patients and health-care professionals as part of patients' diagnosis and management. Usual care reflected the range of ser-vices available in the trial sites, excluding telehealth.

Main outcome measure: Proportion of patients admitted to hospital during a 12-month trial period.

Results: Patient characteristics were similar at baseline. Compared with controls, the intervention group had a lower admission pro-portion within 12-month follow-up (odds ratio 0.82, 95% confidence interval 0.70 to 0.97, $P=0.017$). Mortality at 12 months was also lower for intervention patients than for controls (4.6% v 8.3%; odds ratio 0.54, 0.39 to 0.75, $P<0.001$). These differences in admissions and mortality remained significant after adjustment. The mean number of emergency admissions per head also differed between groups (crude rates, intervention 0.54 v control 0.68); these changes were significant in unadjusted comparisons (incidence rate ratio 0.81, 0.65 to 1.00, $P=0.046$) and after adjusting for a predictive risk score, but not after adjusting for baseline characteristics. Length of hospital stay was shorter for intervention patients than for controls (mean bed days per head 4.87 v 5.68; geometric mean difference −0.64 days, −1.14 to −0.10, $P=0.023$, which remained significant after adjustment). Observed differences in other forms of hospital use, including notional costs, were not significant in general. Differences in emergency admissions were greatest at the beginning of the trial, during which we observed a particularly large increase for the control group.

Conclusions: Telehealth is associated with lower mortality and emergency admission rates. The reasons for the short-term increase in admissions for the control group are not clear, but the trial recruitment processes could have had an effect (Steventon et al., 2014).

Telecare and Its Integration into Traditional Health-Care Services

The health systems still have access barriers to service the greater population. As society ages, greater demands for services will put a strain on the health-care systems and health professionals. Telecare is able to keep individuals safe and independent, offering remote care for the elderly and physically less able people. The main goal is for patients to receive the necessary care and reassurance needed to allow them to remain living in their own homes. Therefore, telecare applications offer a variety of solutions to many health-related issues, particularly for the aging population. The socio-economic challenges of an aging population pose significant challenges to the provision of acute and long-term health care and consequently to SDGs. Just about 28 million persons worldwide suffer from dementia and its financial expenditure is approximately US$156 billion in direct care costs per year (Wimo et al., 2006). According to the U.S. Department of Commerce, U.S. Census Report − 2014, between 2012 and 2050, the United States will experience considerable growth in its older population. In 2050, the population aged 65 and over is projected to be 83.7 million, almost double its estimated population of 43.1 million in 2012 (Ortman et al., 2014). Elderly people may need different assistant technologies to provide them with independence and quality of life; thus, monitoring technology around them would provide with the tools to minimize the safety risks in their homes. In current situations, a caretaker stays in the home with the elderly for supervision and care. Obviously, this action interferes with their privacy and daily interactions; telecare technology can be a solution to this problem.

Figure 4.3 shows the prototyping telecare application implementation to integrate pure-software components and sensors. To make a sensor accessible in the service framework, it is necessary to connect the physical device to a computer and write a wrapper service that communicates with the sensor to obtain the readings. The connection between a sensor and the computer would be wired or wireless, depending on the deployment requirements.

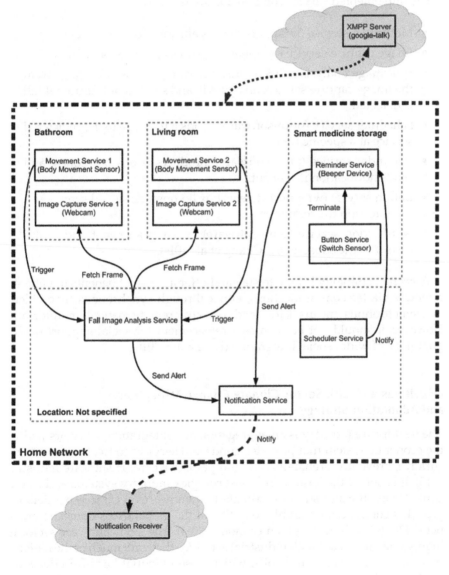

FIGURE 4.3
A framework for prototype telecare applications (Chang et al., 2014).

Here we assume the connections are wired and it is sufficient for demonstration purposes. In the implementation, we connect the sensors to the Arduino Mega 2560 board. The program uploaded to the Arduino board translates the sensor readings to a value recognized by the corresponding wrapper service. The wrapper communicates with the Arduino board through the USB (emulated RS232) port. The software services are implemented with the Python SleekXMPP [21] library.

The implemented service components are list below:

- Movement service (wrapper): detects abrupt body movement.
- Image capture service (wrapper): captures an image on demand.
- Fall image analysis service (pure-software): fetches an image from the image capture service, analyzes it, and sends a notification if fall is detected.
- Scheduler service (pure-software): sends a message to the registered service at a specific time.
- Reminder service (pure-software): produces beeps on request and sends notification if timeout.
- Button service (wrapper): detects box lid open-close and sends messages to stop the reminder.
- Notification service (pure-software): receives an event and multicasts it to the subscribers (Chang et al., 2014).

Telecare technology is here to stay and offer a series of benefits to patients remotely via telecommunications, either through synchronous (live video) or asynchronous means (store-and-forward, remote patient monitoring). However, it should not be seen as a replacement for social support or the guarantee of wider social integration (Roulstone, 2016).

eHealth as a Health Sector Climate Change Mitigation and Adaptation Strategy

The health-care industry is an aggregation and integration of sectors within the economic system that provides goods and services to treat patients with curative, preventive, rehabilitative, and palliative care (Healthcare industry, 2015). It is one of the largest and most complex industry systems and in the United States the industry has estimated expenditures of US $639 per person-year that corresponds roughly to 8–10% of the global gross domestic product (GDP) (Health in the green economy, 2011). The US health-care sector is highly interconnected with industrial activities that emit much of the nation's pollution to air, water, and soils, which pose potential harmful effects on public health. Furthermore, the US health-care industry is the second largest energy consumer among all US industrial sectors, and its in-patient facilities are the second most energy-intensive commercial buildings in the country. The National Health Service (NHS), one of the largest European employers, has a carbon footprint of approximately 20 million tons of carbon dioxide (CO_2) per year (Sustainability in the NHS, Health check, 2012). According to the UK Department of Energy and Climate Change statistical release (2011), it is equivalent to more than 4% of annual emissions; NHS buildings consume energy worth over £410 million (approximately US$556 million) annually (Saving Carbon, Improving Health, NHS carbon reduction strategy for England, 2009). Furthermore, while the percentages are expected to be lower in developing countries, in the developed world, the health-care sector is estimated to contribute 3–8% of the total emissions. For example, when released as atmospheric gases, inhaled anesthetics from health-care facilities have a much higher global warming potential – hundreds to thousands of times higher – than comparable quantities of carbon dioxide. These gases are emitted into the atmosphere because of poor waste management and lack of recycling strategies (Saving Carbon, Improving Health, NHS carbon reduction strategy for England, 2009; Health in the green economy. co-benefits to health of climate change mitigation, 2011; World Health Organization: Safe health-care waste management, 2004).

As a mobile industry, the consumption of fossil fuels in the health sector (when patients and medical professionals travel to and from appointments, pick up prescriptions, and obtain tests and results) is a serious concern because of the implications of the rapidly increasing risk of adverse effects on health from climate change effects (Holmner et al., 2012). Considering the increasingly detrimental effects of severe weather events, exacerbating its effects among poor communities, squatter settlement dwellers, the homeless, and other susceptible groups that are more vulnerable to climate change, mitigation, and adaptation strategies needs to be tailored to the health-care sector planning.

Eckelman and Sherman (2018) quantified the increased disease burden caused by the US health-care sector in their report titled: "Estimated Global Disease Burden from US Health Care Sector Greenhouse Gas Emissions." The study estimated that life cycle greenhouse gas (GHG) emissions associated with US health-care activities will cause an additional 123,000 to 381,000 DALYs (disability-adjusted life years) annually (Figure 6.1) based on 2013 health-care sector life cycle emissions. The study used the same proportions of disease contributions to total DALYs for each health damage factor reported in Tang L, Ii R, Tokimatsu K, Itsubo N. "Development of human health damage factors related to CO_2 emissions" (2015) and De Schryver AM et al. "Characterization factors for global warming in life cycle assessment based on damages to humans and ecosystems" (2009). In all cases, the largest potential for health damages was attributed to malnutrition (49–63% of the total), which will particularly affect regions with large populations and agricultural areas located on floodplains, or regions lacking irrigation including much of Africa and parts of South and Southeast Asia. Increased incidence of diarrhea and malaria are the other main contributors to total DALYs, because of lengthening warm seasons and expanding geographic ranges of disease vectors. Adding the average estimate of 209,000 DALYs to the earlier figure reported by Eckelman and Sherman. Environmental impacts of the US health-care system and effects on public health (2016) for non-GHG results increased US health-care-related public health damages to a total of 614,000 DALYs per year for all emission types. Because actual global GHG emissions exceeded those predicted in the Special Report on Emissions Scenarios (SRES) for the early twenty-first century (Raupach et al., 2007), the health damage factors we used may well underestimate actual health damages over the coming decades.

eHealth for a Sustainable World

The health-care industry needs to boost sustainability efforts and reduce its environmental impact in order to achieve a greener business operation, minimize its carbon footprint, and achieve some other benefits such as institutional financial gain, improved patient outcomes, better staff health and reduced turnover, and community engagement. HIT, also referred to as eHealth, can support the sector's sustainability program and tailor a pre-existing strategy. eHealth alone cannot be the solution to a sustainable health sector. It needs to be accompanied by a comprehensive sustainability strategy and risk management framework. Nevertheless, building foundations for eHealth to deliver public health and health services in a more strategic and integrated manner is essential for the effectiveness of national policies, strategies, and governance to ensure the progress and long-term sustainability of investments. The term eHealth is broad and includes work within a health-care organization where ICT is the central element and foundation. Electronic medical records, home monitoring of vital parameters using mobile technology as well as electronic health-surveillance systems are considered eHealth (Holmner et al., 2012). A strategic framework and policies creating the foundation of an eHealth development plan are fundamental for the implementation of a successful eHealth system. Such e-strategies can help cross implementation barriers and address different views by involving all stakeholders in a common project and focus energy and resources into key development objectives. The proposed initiatives, strategic eHealth planning and policies, should be legislated in such a way as to enable eHealth applications and services, accelerate emissions reduction, increase the quality and efficiency of care, reduce erroneous treatments, improve access to care in remotely populated areas, and help address the realities of a new health-care environment.

Climate Change Mitigation and Adaptation Strategy for the Health Sector

The following strategies published in "Climate change and eHealth: a promising strategy for health-sector mitigation and adaptation" (2012) represent important steps to consider while developing mitigation and adaptation planning frameworks:

1. Building green, which includes strategies to conserve energy (Houghton, 2011; Younger et al., 2008).
2. Efficient energy distribution and use of renewable energy sources (Edward Vine, 2012; Omer, 2008).
3. Passive or low-energy cooling, heating and ventilation strategies (Short et al., 2004).
4. Strategies for conserving and maintaining water resources (Hanak and Lund, 2012; United Nations Educational Scientific and Cultural Organization, 2009).

Other strategies are more specific for the health-care sector and include:

1. Reducing GHG emissions from anesthetic gas use and waste management (Ryana and Nielsen, 2010; Barwise et al., 2011).
2. Increased use of HIT, such as eHealth (Health in the Green Economy, 2011; Yellowlees et al., 2010).

The factors representing the success of eHealth as a mitigation strategy are related to the type of service needed for investment in new equipment and the lifespan of the technology to effectively reduce CO_2 emissions. The environmental impact and costs of ICT technology such as manufacture, distribution, daily use, and subsequent disposal of waste need to be assessed. However, a full carbon-cost benefit analysis for these applications can be a challenge since all of the factors contributing to telemedicine's carbon footprint are not adequately studied. Nevertheless, there are many programs addressing green and environmental computing or "green ICT" including greener manufacturing of components, increased energy efficiency, and enabling more efficient use of existing technology, leading to the use of eHealth as a strategic tool to the reduction of carbon emissions (WWF Sweden, 2008; Forge et al., 2009).

Telemedicine has the potential to mitigate carbon emissions and, consequently, reduce air pollution, known to be adversely associated with disease and deaths, by reducing travel and transportation. Though not all telemedicine applications will reduce travel, the benefits are obvious for home care programs and outpatient consultations. The telemedicine program at

University of California – Davis, is one of the largest telehealth programs in the country, expanding its reach through the center's internationally recognized and accredited education program, providing leadership in the creation of the California Telehealth Network (CTN). Thus far, it has involved 13,000 outpatient consultations over a period of five years, and it has resulted in a savings of 4.7 million miles of travel and a reduction of 1,700 tons of CO_2 emissions (Yellowlees et al., 2010). In Canada, it has been estimated that more than 11 million home visits by nurses could be replaced by telecare, which would result in a reduction of about 75 million miles of travel and about 36.62 tons of associated GHG emissions annually (Scott et al., 2009). Therefore, the reduction in carbon emissions, travel time, and cost-effective operations show the benefits and potentials of eHealth as a mitigation strategy.

The potential for carbon reduction emissions will be determined by several factors related to the number of users and appointments that can be replaced by virtual visits. The benefits will depend on the distance and type of transportation replaced by technology, related to travel by car, public transportation, or airplane, and the distances between health-care professionals and the patients involved.

The reduction potential is dependent on the number of visits as well as the carbon emission caused by each user's travel and visit in a traditional care scenario. The climate impact from travel depends heavily on the type of transportation (e.g., public transportation, car, or airplane), but for simplicity is illustrated as travel distance only. This simplified model does not take into account that each piece of equipment can only serve a limited number of users and visits (Holmner et al., 2012).

eHealth, Public Health, and Environmental Impacts

Potential environmental benefits of eHealth are contingent on the health sector infrastructure and local need. Nevertheless, eHealth has an important and relatively unexplored potential as a health-sector mitigation strategy. eHealth's full potentials are in the first stage of development and execution.

List of Terms

Anesthetic: A substance that induces insensitivity to pain.
Arduino Mega 2560: A microcontroller board based on the ATmega2560. It has 54 digital input/output pins (of which 15 can be used as PWM outputs), 16 analog inputs, 4 UARTs (hardware serial ports), a 16 MHz crystal oscillator, a USB connection, a power jack, an ICSP header, and a reset button.

Digital health: The convergence of the Digital and Genomic Revolutions with health, health care, living, and society. As we are seeing and experiencing, digital health is empowering us to better track, manage, and improve our own and our family's health, live better, more productive lives, and improve society.

Disease burden: The impact of a health problem as measured by financial cost, mortality, morbidity, or other indicators. It is often quantified in terms of quality-adjusted life years (QALYs) or disability-adjusted life years (DALYs), both of which quantify the number of years lost due to disease (YLDs).

Epidemiology: A branch of medical science that deals with the incidence, distribution, and control of disease in a population.

Glycated hemoglobin: A form of hemoglobin that is measured primarily to identify the three-month average plasma glucose concentration.

Health information technology (HIT): Information technology applied to health and health care. It supports health information management across computerized systems and the secure exchange of health information between consumers, providers, payers, and quality monitors.

Hemoglobin: A conjugated protein, consisting of heme and the protein globin that gives red blood cells their characteristic color. It combines reversibly with oxygen and is thus very important in the transportation of oxygen to tissues.

mHealth (mobile health): A general term for the use of mobile phones and other wireless technology in medical care. The most common application of mHealth is the use of mobile phones and communication devices to educate consumers about preventive health-care services.

Post-traumatic stress disorder (PTSD): A mental health condition that's triggered by a terrifying event – either experiencing it or witnessing it. Symptoms may include flashbacks, nightmares, and severe anxiety, as well as uncontrollable thoughts about the event.

Sequelae: Pathological conditions resulting from a prior disease, injury, or attack.

Telecare: The term that relates to technology that enables patients to maintain their independence and safety while remaining in their own homes.

Telehealth: The collection of means or methods for enhancing health care, public health, and health education delivery and support using telecommunications technologies.

Telemedicine: The remote delivery of health-care services, such as health assessments or consultations, over the telecommunications infrastructure.

5

Intelligent Transport Systems for Sustainable Development

The Future of Mobility: Urban Transformation

The global transportation industry is on the verge of undergoing an unprecedented transformation to a new mobility ecosystem. The dynamic mobility transformation could have far-reaching implications for how we travel, when and where one pleases, and of the viability of the choice not to travel. This new revolution is driven by social trends and the driverless vehicles, enabling new business models and mobility services for new and changing customers. Organizations around the world are preparing now for the mobility revolution and transforming operations, implementing new technologies, and refocusing assets and talent to be well positioned to efficiently deliver services in the new personal mobility landscape.

Connected Private and Public Transportation

Each day around the world motorists drive 1.2 billion personal motor vehicles (PMVs). The United States leads the mark in personal motor mobility; daily, motorist's record 13 billion vehicle kilometers traveled (VKT), or 8 billion vehicle miles traveled (VMT). Although the United States only counts for approximately 4.4% of the world's population, American motorists own one-fifth of the world's automobiles, and account for one-quarter of the world's travel by PMV (Schiller and Kenworthy, 2018). Cars are projected to reach the 2 billion mark by 2040, while air travel miles are set to hit 12.42 trillion in the same period. Bernstein said it expects most of this transport growth to happen in emerging markets like China and India, as global populations are set to rise by another 2 billion over the next 25 years to 9.2 billion. The growth in population needs to be supported by a modern, efficient, and reliable infrastructure equipped to support more vehicles, more roads, and more miles traveled. The consequences of more vehicles will result in congestion, more fuel consumption, pollution, and emissions. Nonetheless, rail capacity is going to rise and by 2030, approximately 62,000 miles (100,000 kilometers) of additional rail track needs to be built to support the capacity and to ensure trains run safely at optimum track speed (Global Land Transport Infrastructure Requirements, 2013). The SMARTer2030 (2015) report highlighted the inefficiencies of global logistics operations with unused or excess capacity across supply chains. These inefficiencies will exacerbate as the number of people and products transported increases, creating fuel, energy, and materials no longer useful for production.

Sustainable Mobility and Logistics

Sustainability in the context of a natural resource policy would mean limiting depletion of resources to the rate at which they can be replenished, or alternatives can be identified (Goldman and Gorham, 2006). Implementing sustainability as the core principle of action and behaviors within organizations can potentially have far-reaching implications for policy and for any effective attempt to stimulate "leading change." Therefore, technologies and innovations that promise to advance sustainability efforts can have a positive impact related to the sustainable development goals of economic development, social and human development, and environmental and ecological health. The aforementioned goals can be achieved with an enhanced connectivity through ICT and could significantly increase efficiency while reducing congestion, emissions, and resource consumption, which could make personal mobility more sustainable and efficient (SMARTer2030, 2015). Innovative technologies like electric vehicles and driver-less transportation, in addition to the emergence of new business models for providing shared mobility services such as ride-sourcing or "transportation network companies," (e.g., Lyft, Uber), ride-splitting (e.g., UberPOOL, Lyft Line), and ridesharing (e.g., carpooling, Scoop, Waze) are rapidly redefining mobility and changing the way people move. The Energy Department National Laboratories reported a broad range of energy impacts from automated and connected vehicles. The conclusions highlighted that the impact on energy could range from a possible 60% system-wide decrease in energy consumption to a 200% system-wide increase in energy consumption* (The Transforming Mobility Ecosystem: Enabling in Energy-Efficient Future, 2017). Energy-efficient mobility ecosystems are intrinsically connected to technological and policy initiatives to promote sustainable development, reducing energy consumption, and harness sustainability. Hence, a sustainable mobility needs to create benefits such as increased safety, accessibility, reliability, and affordability, while at the same time reducing energy consumption in the transportation system for an energy-efficient mobility future (The Transforming Mobility Ecosystem: Enabling in Energy-Efficient Future, 2017) (Figure 5.1).

---

* Study assumes continued use of internal combustion engines with no fuel or powertrain switching other than potential downsizing of conventional powertrains. For a full set of assumptions and methodology, see Stephens et al. (2016). Estimated Bounds and Important Factors for Fuel Use and Consumer Costs of Connected and Automated Vehicles." National Renewable Energy Lab.

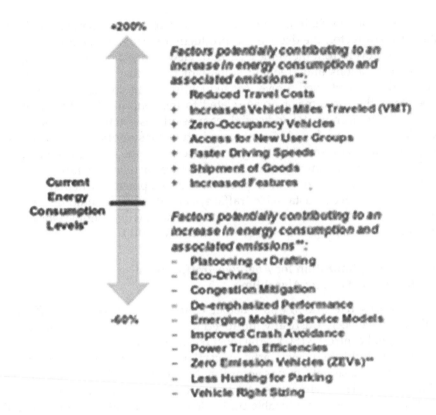

FIGURE 5.1

Energy impacts of connectivity and automation (The Transforming Mobility Ecosystem: Enabling in Energy-Efficient Future, 2017).

*For assumptions and methodology, see Stephens et al. (2016). "Estimated Bounds and Important Factors for Fuel Use and Consumer Costs of Connected and Automated Vehicles." National Renewable Energy Lab. November. **Zero Emission Vehicles (ZEVs) not accounted in research; fuel switching could increase efficiency gains.

Traffic Control and Optimization

Traffic control and optimization become a significant aspect of efficient operations considering population growth, more frequent travel, and traffic congestion that leads to travel delays. The consequences of these actions create waste in fuel consumption, affects work productivity that increases both stress and carbon emissions. There are several strategies and tools to reduce excessive fuel consumption and vehicular emissions on urban streets. The optimization of signal timing tools can be used to reduce traffic delays and stops. Rising traffic congestion is an inescapable condition in large and growing metropolitan areas across the world and inadequate transport infrastructure, road and highway facilities, and ineffective means for the supply and distribution of goods are connected to high delay times for important transportation which, in a wider sense, can affect the economic growth. In the business sector, the decrease in productivity leads to increases in labor costs. Moreover, from the vehicle point of view, congestion results in higher speed fluctuations and frequent stops which increase the fuel consumption and, as mentioned above, consequently results in higher emissions. Therefore, the three major areas related to the concept of sustainability – economic, environmental and social responsibilities – in the business sector, are detrimentally affected. An increase in fuel consumption and emissions impacts our environment by the greenhouse effect, the social health by pollution and our economy by increased fuel prices. Sustainable traffic systems are an imperative component of sustainable development, particularly in cities, and to achieve a sustainable traffic system, the focus should be on both mobility and environment (Optimal Signal Control with Multiple Objectives in Traffic Mobility and Impacts, 2009; Kwak et al., 2012).

Real-Time Optimization for Adaptive Traffic Signal Control Using Genetic Algorithms

The advances in technology have helped control methodologies of traffic signals that have been drastically improved in the last few years. The adaptive traffic signal control (ATSC) is the most recent and advanced control type of traffic signal; able to relieve traffic congestion in an efficient way by continuously adjusting signal timings according to real-time traffic conditions (Lee et al., 2005). ATSC technologies are best suited for arterials that experience highly variable or unpredictable traffic demands for which multiple signal timing solutions are necessary during a typical time-of-day period. Many studies have shown ATSC improves average performance metrics (travel time, control delay, emissions, and fuel consumption) by 10% or more and in outdated systems and working under saturated conditions, these benefits can considerably be increased by five times the average performance metrics. In areas where traffic demand is stable and predictable during typical time-of-day periods, performance is regularly monitored, and signal timing is well maintained these improvements are not as significant, but the overall improvement of the system performance is worth notated (Adaptive Signal Control Technologies, 2016).

The methods used for traffic signal operation such as integer programming, hill climbing, or descent gradient searching were gradually surpassed by genetic algorithms (Gas) (Zhou Guangwei, 2006).

There are a number of specific characteristics of GAs. These characteristics provide an advantage over other methods of optimization:

- A GA works from a population, not a single point, and hence it is less likely to be trapped at a local optimum.
- Derivative freeness, i.e., a GA does not need the objective function's derivative to do its work.

- Flexibility, i.e., a GA can function just fine regardless of how complex the objective function is; the only thing it requires of the function is that it is executable (i.e., its value can be calculated given the values of the decision variables).
- Because of its implicit parallelism, a GA can handle combinatorial problems efficiently. It has been shown that as the size of the search space or number of solutions increases exponentially, the time requirements for the GA to reach solutions only grow linearly. This feature is particularly useful for online optimization of transportation problems such as traffic control.
- A GA naturally lends itself to parallel implementation. This follows from its functional components structure.
- GA is, for the most part, based on intuitive notions and concepts (Turky et al., 2009).

Intelligent Transportation Systems for Improving Traffic Energy Efficiency and Reducing Greenhouse Gas Emissions from Roadways

The U.S. Department of Transportation (USDOT) conducted a report in 2014 about how the field of Intelligent Transportation Systems (ITS) increased activity in recent years, with the application of modern control, communications, and information technologies to vehicles and roadway infrastructure. ITS deliver efficiency and improvements in transportation system performance with emphasis on travelers' safety and convenience and reducing congestion. The ITS can be categorized into three major areas: Vehicle Systems, Traffic Management Systems, and Travel Information Systems (TIS).

Vehicle Systems

Modern control systems, faster on-board processors, and wireless communications provide features that greatly improve vehicles performance. The development of traveler information services, the integration of different types of sensors, tracking systems, and mobile communications technologies are moving into the Internet of Things (IoT). Real-time communications will connect the components of cars together into intelligent machines. Vehicles with modern control systems are increasingly becoming intelligent infrastructures that are safer and more efficient. Emerging vehicle systems are exemplified by the following (Figure 5.2):

Longitudinal assistance system functions include adaptive cruise control and forward collision warning/avoidance (Wang et al., 2012). On-board radar, LiDAR, and computer vision technology monitor headways between vehicles, and by providing feedback to the vehicle's braking system, integrating safety controls and consequently, reducing collisions (Barth et al., 2015). These sensors are also being used for Adaptive Cruise Control (ACC)

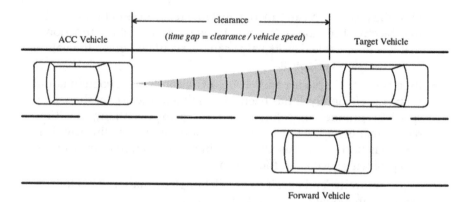

FIGURE 5.2
Longitudinal assistance systems (Wang, 2012).

systems, which allow a vehicle's cruise control system to adapt the vehicle's speed to the traffic environment. A radar system attached to the front of the vehicle is used to detect whether slower moving vehicles are in the ACC vehicle's path (Intelligent Transportation Systems for Improving Traffic Energy Efficiency and Reducing GHG Emissions from Roadways, 2015).

The following shows the definitions for Figure 5.2, representing the physical overview and functions of the system:

- *Adaptive Cruise Control (ACC)*: An enhancement to a conventional cruise control system which allows the ACC vehicle to follow a forward vehicle at an appropriate distance.
- *ACC vehicle*: The subject vehicle equipped with the ACC system.
- *Active brake control*: A function which causes application of the brakes without driver application of the brake pedal.
- *Clearance*: Distance from the forward vehicle's trailing surface to the ACC vehicle's leading surface.
- *Forward vehicle*: Any one of the vehicles in front of and moving in the same direction and traveling on the same roadway as the ACC vehicle.
- *Set speed*: The desired cruise control travel speed set by the driver and is the maximum desired speed of the vehicle while under ACC control.
- *System states: ACC off state*: Direct access to the "ACC active" state is disabled.
- *ACC standby state*: System is ready for activation by the driver.
- *ACC active state*: The ACC system is in active control of the vehicle's speed.

- *ACC speed control state*: A substate of "ACC active" state in which no forward vehicles are present, such that the ACC system is controlling vehicle speed to the "set speed" as is typical with conventional cruise control systems.
- *ACC time gap control state*: A substate of "ACC active" state in which time gap, or headway, between the ACC vehicle and the target vehicle is being controlled.
- *Target vehicle*: One of the forward vehicles in the path of the ACC vehicle that is closest to the ACC vehicle.
- *Time gap*: The time interval between the ACC vehicle and the target vehicle. The "time gap" is related to the "clearance" and vehicle speed by: time gap = clearance/ACC vehicle speed (Adaptive Cruise Control System Overview, 2005; Plessen et al., 2018) (Figure 5.3).

A Cooperative Adaptive Cruise Control (CACC) system has been developed by adding a wireless vehicle-vehicle communication system and new control logic to an existing commercially available ACC system (Nowakowski et al., 2010). The CACC includes multiple concepts of a communication-enabled vehicle following and speed control, improving the vehicle-following capabilities of ACC. Therefore, CACC is an area of increased development where vehicles communicate with each other to "cooperatively" manage traffic control; to have a vehicle's cruise control system maintain a proper following distance behind another car by slowing down once it gets too close, braking to achieve safety, and minimizing impacts and other risks.

Changing lanes always poses a risk, especially if a vehicle in the neighboring lane is accelerating. There is also the "blind spot" to consider – an area located behind the car that is not visible in the external rear-view mirrors. *Lateral Assistance Systems* are designed to improve the performance of vehicles during lane changes, merges, or any kind of turning movement. Computer vision technology and other sensor technologies coupled with wireless communications are being deployed to provide lane departure warnings and to

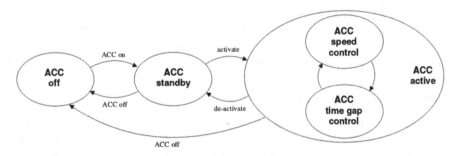

FIGURE 5.3
ACC states and transitions (Adaptive Cruise Control System Overview, 2005).

advise drivers of pending lateral collisions (Barth et al., 2015). Consequently, autonomous navigation for intelligent vehicles usually requires a control system for path following. This control system is composed of longitudinal and lateral controllers. The longitudinal controller is responsible for regulating the vehicle's cruise velocity while the lateral controller steers the vehicle's wheels for path tracking (Filho et al., 2014). Longitudinal and lateral control systems are now leading us down the path toward partial and full automation, and it promises new levels of efficiency and minimal risks. With the growth of wireless communications, such as Wi-Fi, WiMax, cellular-based information delivery, and satellite radio, the vehicle communications systems have been enhanced and deployed in many different ways. Cellular communication technology is already playing a large role in fleet management applications and vehicle monitoring. Nevertheless, Dedicated Short Range Communication (DSRC) radios are a two-way short- to medium-range wireless communications capability that permits very high data transmission critical in communications-based active safety applications (DSRC: The Future of Safer Driving. Intelligent Transportation Systems Joint Program Office, 2018) and will likely be deployed to enable vehicle-to-vehicle (V2V), vehicle-to-infrastructure (V2I), and infrastructure-to-vehicle (I2V) applications. These features are mainly utilized in safety measures (Barth et al., 2015). Additionally, mobility and environmental applications that support a more sustainable relationship between transportation and the environment, primarily through fuel use and subsequently emissions reductions benefits, are emerging as an important tool to develop and implement effective ways to sustainability (Connected Vehicles – Environmental Applications, 2017).

The U.S. Department of Transportation's (USDOT's) Connected Vehicle program tests and evaluates technology that will enable cars, buses, trucks, trains, roads and other infrastructure, and our smartphones and other devices to "talk" to one another. Therefore, applications that enable tools to minimize environmental risks are being deployed. Employing a multimodal approach, the AERIS Research Program supports the development of technologies and applications that enhance sustainability strategies in the transportation sector. The AERIS Capstone report (2009–2014), summarizes the five years of connected vehicle environment research conducted by the Intelligent Transportation Systems (ITS) Joint Program Office (JPO) and presents a comprehensive modeling benefit for the environmental applications that were being researched (Table 5.1).

Traffic Management Systems

The total amount of driving, as measured in VMT, has increased significantly, creating a number of issues such as congestion, accidents, pollution, etc., in many urban areas around the world. Modern infrastructure is one of the solutions to cope with the aforementioned issues, but it is not always feasible, depending on budgets, regulations, urban design challenges, geographic and environmental

TABLE 5.1

Summary of AERIS Modeling Results

Eco-Approach and Departure at Signalized Intersections	• The application provided *5–10% fuel reduction benefits* for an uncoordinated corridor.
	• For a coordinated corridor, the application provided up to *13% fuel reduction benefits*.
	   • 8% of the benefits were attributable to signal coordination.
	   • 5% attributable to the application.

Eco-Traffic Signal Timing	• When applied to a signalized corridor that was fairly well optimized, the application provided an *additional 5% fuel reduction benefit* at full connected vehicle penetration.
Eco-Traffic Signal Priority	• The Eco-Transit Signal Priority application provided *up to 2% fuel reduction benefits* for transit vehicles. • The Eco-Freight Signal Priority application provided up to *4% fuel reduction benefits* for freight vehicles.
Connected Eco-Driving	• When implemented along with a signalized corridor, the application provided *up to 2% fuel reduction benefits* at full connected vehicle penetration. • The application provided *up to 2% dis-benefit in mobility* (e.g., travel time) due to smoother and slower accelerations to meet environmental optimums.
Combined Eco-Signal Operations Modeling	• Together the Eco-Signal Operations applications provided *up to 11% improvement* in CO_2 and fuel consumption reductions at full connected vehicle penetration
Eco-Speed Harmonization	• The application provided *up to 4.5% fuel reduction benefits* for a freeway corridor. It assisted in maintaining the flow of traffic, reducing unnecessary stops and starts, and maintaining consistent speeds near bottleneck and other disturbance areas.
Eco-Cooperative Adaptive Cruise Control (Eco-CACC)	• Eco-CACC provided *up to 19% fuel savings* on a real-world freeway. • Vehicles using a dedicated "eco-lane" experienced 7% *more fuel savings* when compared to vehicles in the general lanes. • Eco-CACC has the potential to provide *up to 42% travel time savings* on a real-world freeway corridor for all vehicles.
Combined Eco-Lanes Modeling	• Together the Eco-Lanes applications provided *up to 22% fuel savings* on a real-world freeway corridor for all vehicles. • Vehicles using the dedicated "eco-lane" experienced 2% *more fuel savings* when compared to vehicles in the general traffic lanes. • The scenario provided *up to 33% travel time savings* for all vehicles.
Low Emissions Zones	• A Low Emissions Zone modeled in the Phoenix Metropolitan Area resulted in up to *4.5% reduction in fuel consumption* when both eco-vehicle incentives and transit incentives were offered. • The modeling indicated that the Low Emissions Zone has the potential *to reduce VMT by up to 2.5%* and *increase by up to 20%* into the Low Emissions Zones.

Source: Connected Vehicles – Environmental Applications, 2017.

factors in place, and more. However, a number of ITS-based traffic management system solutions can help reducing congestion and other hazards.

Traffic Monitoring Systems (TMS) integrate multiple technologies to improve the flow of vehicle traffic and improve safety. TMS use better sensor technology, more reliable communication channels, and more advanced information processing capability. One of the key enablers for having smooth traffic flows and better mobility is to rely on real-time traffic monitoring systems, adding new data processing techniques which are being developed to estimate traffic flow, density, and speed, as well as other microscopic traffic parameters. Hence, real-time traffic information reduces congestion, minimizing accidents in addition to individual drivers choosing alternative routes, saving fuel, and offering time-management efficiency travel.

Improving transportation by taking all elements in a corridor – highways, arterial roads, and transit systems – and consequently optimizing the use of existing infrastructure assets is essential to keep traffic flowing. *Integrated Corridor Management techniques*, such as innovative ramp metering for freeway access ramps, and advanced signal timing algorithms on arterial networks, improve issues with traffic congestion and reduces vehicle's engine running while the vehicle is not in motion, stopping emitted emissions and helping the environment (Barth et al., 2015).

Travel Demand Management is another critical element of traffic management. Helping people know and use all their transportation options to optimize all modes in the system – and to counterbalance the incentives to drive that are so prevalent in urban areas is a significant step to reducing the number of vehicles and, subsequently, congestion or spreading out the peak of traffic volume. Traffic congestion is an epochal problem in urban areas in the United States and around the world. Previous analyses have estimated the economic costs of congestion, related to fuel and time wasted; however, few have quantified the public health impacts or determined how these impacts compare in magnitude to the economic costs (Levy et al., 2010). Reducing greenhouse gas (GHG) emissions is crucial to sustaining healthy environments and keeping traffic flowing smoothly at moderate speeds will have a large impact on reducing energy consumption and GHG emissions. Therefore, it is necessary to highlight the importance of conducting assessments accounting for congestion in emission, exposure and health risks associated with traffic management.

Travel Information Systems – Information systems for travelers are significantly evolving and provide *travelers* with information that will facilitate their decisions concerning route choice, departure time, trip delay or elimination, and mode of transportation (Kristof et al., 2005).

Examples of this technology include the following:

> *Route guidance systems* are one of the main components of travel and transportation management in vehicular roadways and a tool to alleviate congestion. The market growth of low-cost electronics such

as sensors, wireless communication, and computing equipment has now made large-scale vehicle navigation available to consumers. It is a system that considers driver preferences, vehicle parameters such as speed and performance capabilities in a navigation computer, and outputs flexible guidance instructions. Alleviating congestion is a significant task considering the growth in population density within cities in the urban transportation network.

These navigation systems include on-board, off-board, and smart-phone-based systems and use geographic and real-time traffic information. The growth of road traffic and the increasing inconvenience and environmental damage caused by road congestion require a significantly more efficient use of the infrastructure. Therefore, selecting optimal routes in a roadway network from specific origins to specific destinations minimize vehicle volume on the roads, health risks caused by emissions, travel time, and travel distance.

Geo-location systems are typically coupled with route guidance systems to allow users to find specific locations, cutting down on excessive driving (e.g., searching for a gasoline filling station, open parking space, etc.).

Electronic payment systems eliminate the need for a driver to decelerate the vehicle, idle while a manual transaction takes place, and then accelerate the vehicle back to the desired speed, adding convenience for the traveler while reducing GHG emissions (Barth et al., 2015).

List of Terms

Adaptive cruise control (ACC): A driver assistance technology that sets a maximum speed for vehicles and automatically slows the speed of the car when traffic is sensed in front of the vehicle.

Adaptive traffic signal control (ATSC): A traffic management strategy in which traffic signal timings change, or adapt, based on observed traffic demand.

Cooperative Adaptive Cruise Control (CACC): An extension to the adaptive cruise control concept. CACC realizes longitudinal automated vehicle control.

Genetic algorithm (GA): A heuristic search method used in artificial intelligence and computing. It is used for finding optimized solutions to search problems based on the theory of natural selection and evolutionary biology.

Integrated Corridor Management (ICM): An approach to improving transportation by taking all elements in a corridor, including highways, arterial roads, and transit systems into account.

Intelligent transportation system (ITS): The application of sensing, analysis, control and communications technologies to ground **transportation** in order to improve safety, mobility, and efficiency.

Sustainability: In environmental science, is the quality of not being harmful to the environment or depleting natural resources, and thereby supporting long-term ecological balance.

Travel Information Systems (TIS): Systems that are designed to assist travelers in making better travel choices by providing information regarding the available travel alternatives.

6

ICT Role and Roadmap for Smart Sustainable Cities

The Role of Information and Communication Technology in Building Smart Cities

A city is a system of systems. This model applies to the dependency of the city's resilience to its systems such as transportation, utilities, and public health infrastructure and communication networks. These systems are individually dependent and also rely on how interdependent those systems are between them. As rapid urbanization is presenting new human and environmental outcomes, we are presently witnessing that cities are the future of humankind (Marolla, 2017). The challenges for cities in the twenty-first century are numerous as the global community faces the threats of climatic change, dense urbanization, cities' population growing at an accelerating pace, environmental degradation and loss of habitats and ecological connectivity, and rising impacts to people and property due to natural disasters. According to the International Energy Agency (IEA), urban areas now account for more than 71% of energy-related global greenhouse gases (GHGs). This percentage, which is linked to the growth of urbanization and population density, will increase to 76% by 2030. Thus, energy-related emissions are the largest single source of GHGs when looking at allocated allowances for the areas in question (Marolla, 2017; Hoornweg et al., 2011). The design of the built environment can potentially shape the well-being of future generations with a strategic planning and engineering of environmentally sound solutions. Managing and minimizing risks provides capital investment opportunities for cities through initiatives undertaken by local governments and with the involvement of the private sector. Capital investment must focus on upgrading the city's infrastructure through deep energy retrofits by involving owners and investors in carrying them out across the city's infrastructure – supported by a strategic-minded finance community with a focus on the fundamental facilities and systems serving the city, including the services and facilities necessary for its economic function (Marolla, 2017). The concept of "smart cities" is rooted in the creation

and connection of human capital, social capital, and information and communication technologies (ICTs) infrastructure to generate greater and more
sustainable economic development and a better quality of life (European
Parliament, 2014) and is dominating the conversation around the future of
urban environments. Smart cities are part of a broader transition to the digital economy. ICTs enhance the quality and performance of urban services,
minimizing environmental impact and resource consumption, and engaging more actively with its citizens.

The potential benefits of a smart city can be achieved with Public-Private
Partnership (PPP) engagements. According to the World Bank, PPPs are in
general medium- to long-term arrangements between the public and the private sector with clear agreements on shared objectives for the delivery of
public infrastructure and/or public services, in which the private party bears
significant risk and management responsibility, and remuneration is linked
to performance (The World Bank: "What Are Public Private Partnerships?",
2018). The vision to materialize the development of smart cities combines a
long-term view in investment and the collaboration of different sectors in
society, along with a number of converging factors that can turn a munici-
pality's vision for a smart city into reality. Smart cities' developments are difficult to fund with traditional public finance. In this context, PPP presents a
potential solution to overcome the shortage of public finance and budget cuts
on public spending. In summary, some of the smart city benefits are:

- Faster reaction to public safety threats by real-time analysis of sensor and surveillance camera video data;
- Smoothened distribution of tourists (geographical and over time) by
  analysis of tourist movements and real-time incentives;
- Exchange of products and services in a peer-to-peer model (sharing
  economy, from possession to use);
- Dynamic groups of citizens organize themselves to work together
  on collective interests;
- Co-creation of decision-making, new forms of digital democracy
  and participatory government;
- Data-driven policy making leads to more focused interventions and
  measured evidence of effectiveness;
- Better diagnostics and personalized treatment through artificial
  intelligence on massive volumes of patient data;
- People who need care can live in their own home longer through
  advanced sensoring and health-care robotics;
- Lower congestion and pollution through the optimal use of transportation infrastructure (roads, parking places);
- Energy savings through real-time insight in energy usage, combined
  with gamification concepts;

- Responsive household appliances react to dynamic energy prices to adjust energy demand to supply;
- More efficient waste collection due to sensors in waste containers;
- Analysis of data provided by sensors in the water distribution network identifies leakages and allow fast repairs (Smart Cities: How Rapid Advances in Technology Are Reshaping Our Economy and Society, 2015).

Smart City Standardization

The British Standardization Institute (BSI) defines a standard as, "A document established by consensus and approved by a recognized body that provides, for common and repeated use, rules, guidelines, or characteristics for activities or their results, aimed at the achievement of the optimum degree of order in a given context." Standards should be based on the consolidated results of science, technology, and experience aimed at the promotion of optimum community benefits (BSI: A standards for standards, 2011). The International Organization for Standardization (ISO) is a worldwide federation of national standards bodies with 119 members. The results of ISO's work are published in the form of international standards. Sustainable urban development is necessary to achieve quality of life. Strategies that are broader in vision and more regional in scale can be developed using international standards because they deliver efficiencies in laying down requirements, specifications, frameworks, or characteristics that can be used systematically to ensure that materials, products, processes, and services are fit for their purpose and support the established strategy, providing a model to follow when setting up and operating a management system (Marolla, 2017; Temmerman, 2000).

There are significant improvements using standards for smart cities as well as helping city leaders set tangible targets, including service quality and quality of life. Some of the benefits associated with international standards are:

- *Environmental management*: Use fewer resources, cut waste, increase recycling, and reduce landfill.
- *Energy management*: Reduce your energy costs.
- *Emissions verification*: Don't just declare it, verify it.
- *Carbon neutrality*: Achieve net zero carbon emissions.
- *Social accountability*: Understand and improve your impact on your local community.
- *Sustainable procurement*: Show your commitment to renewable resources.
- And much more, from sustainable communities to corporate social responsibility (Becoming More Sustainable with Standards, 2018).

Projections show that urbanization combined with the overall growth of the world's population could add another 2.5 billion people to urban populations by 2050, with close to 90% of the increase concentrated in Asia and Africa (UN DESA'S Population Division, 2015) (Figure 6.1).

As the world population is increasing, dwellers are moving from a rural to an urban area for different factors, but mostly driven by economic and employment reasons. Cities need to make better use of resources and become more efficient. Sustainable development needs to be supported by policies, regulation, citizen involvement, and standards that are all key components needed to build a viable smart city. Besides technology advancements and the role of ICTs to build a viable sustainable society in a path toward smarter cities standardization will play a key role in ensuring consistent outcomes. The UN World Economic and Social Survey 2013 suggested Africa, Asia, and

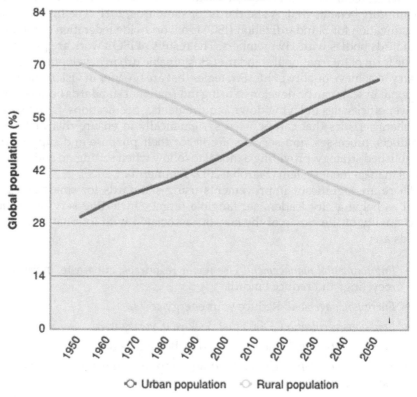

FIGURE 6.1
Global proportion of the urban and rural population (World Health Organization, 2016). The figure shows the urban population in 2015 accounted for 54% of the total global population, up from 30% in 1950 – it is expected to increase to 60% of the world population by 2030.

other developing regions will be housing an estimated 80% of the world's urban population in the years to come. For example, in the period from 1950 to 2010, small cities saw a net increase of 1.3 billion people, while medium and large cities saw an increase of 632 million and 570 million, respectively (Sustainable Development Challenges, 2013).

It is clear that the course of human history has been marked by a process of accelerated urbanization. Urbanization can bring many benefits to society and its inhabitants, but if poorly managed will only amplify existing challenges. The gross domestic growth (GDP) may create conditions that organically drive migration from rural to urban areas, but the assumption that urbanization will necessarily drive strong economic growth may be false (Mingxing et al., 2014); that is the main reason to integrate pragmatic solutions – to increase efforts to make cities more sustainable. The urban shift over time has led to the emergence of the *megacity*. Despite the economic success of megacities, their leaders are still seeking solutions to cope with socio-economic inequalities and the rise of the "urban poor." Global risks such as climate change, scarcity of natural resources, and environmental hazards will compound existing vulnerabilities, increasing poverty levels, and primarily affecting the urban poor because they typically live in the most vulnerable areas of a city and are at high risk from the effects of climate vulnerability and change, social and economic disruptions, and other disasters. Overcrowded living conditions, inaccessibility to safe infrastructure, and poor health conditions make the urban poor highly susceptible to the aforesaid impacts (Marolla, 2017; Baker, 2011).

As a result of accelerated urbanization, urban areas are getting more and more congested. This trend in population growth is expected to continue as it brings socio-economic advantages, but it also brings challenges. Rapid urbanization adds pressure to the resource base and increases demand for energy, water, and sanitation, as well as for public services, education, and health care, which puts stress on the city's available resources. Accordingly, social, economic, and environmental issues have become pressing and are interrelated. Cities greatly contribute to environmental degradation on local, regional, and global scales. For example, cities consume as much as 80% of energy production worldwide and account for a roughly equal share of global GHG emissions (Hoornweg et al., 2011). The International Energy Agency (IEA) estimates that urban areas currently account for over 67% of energy-related global GHGs, which is expected to rise to 74% by 2030. It is estimated that 89% of the increase in CO_2 from energy use will be from developing countries (IEA, 2008).

The awareness of risk management has been emphasized considerably in the government and the private sector. The devastating consequences present a challenge to public policy if meaningful actions are not taken into consideration for densely populated cities. Large-scale disasters include Hurricane Sandy, which had overwhelming impacts in October 2012. The hurricane affected 24 states, including the entire eastern seaboard from Florida to New

York, with particularly severe damage in New Jersey and New York. The impacts of natural hazards (e.g., severe weather events, flooding, extreme heat, emergency diseases) and man-made impacts (hazardous materials, cyber-attacks, power service disruptions and blackouts, chemical threat and biological weapons) are a "global shock" that allows for a unique outline of risks. These types of global shocks stream a series of risks that grow to be active threats because they extend across global systems. These threats expand along the urban ecosystem arising in health, environmental, social, or financial risks (Marolla, 2017). Smart cities with the appropriate risk management framework in place should be able to minimize detrimental events; they could help communities adapt to these changes and deliver sustainable and integrated urban development to enable long-term economic growth.

In essence, risk management standards are relevant in the physical world, where they allow for the interconnection of hardware and technologies, but also in the virtual space, where they facilitate data collection/sharing as well as a city's operational efficiencies (Significant Milestone for Smart City Development – Standards Organizations Agree to Work Together to Move Cities to Greater Smartness, 2017).

Artificial Intelligence Competitive Advantage for Sustainable Development

The rapid development of artificial intelligence (AI) provides innovative ways to support and achieve sustainable development goals and accelerates the global dialog that aims to chart a course for computers to benefit all of humanity. AI is quickly becoming a top solution strategy to solve a variety of issues for industry and governments. According to the International Data Corporation (IDC), widespread adoption of cognitive systems and artificial intelligence across a broad range of industries will drive worldwide revenues from nearly US$8 billion in 2016 to more than US$47 billion in 2020 (Cisco Global Cloud Index: Forecast and Methodology 2016–2021, 2018). It is important to recognize that technology developments must be accompanied by tangible solutions to achieve SDGs. Combining satellite imagery and machine learning to predict poverty is a project proposed by a team at the Sustainability and Artificial Intelligence Lab at Stanford University to track progress toward the distribution of poverty. This project aims at more frequent and reliable data on the distribution of poverty than traditional data collection methods can provide. The Stanford team proposes the combination of machine learning with high-resolution satellite imagery, creating more accurate predictions and data on socio-economic indicators of poverty and wealth (Sustainability and Artificial Intelligence Lab, 2018).

Artificial intelligence is beyond machine learning as its collective form is evolving, recognizing, and responding to data flows using algorithms and rule-based logic. This enables cognitive/AI systems to automate a broad range of functions across many industries, potentially improving business processes, modeling, and forecasts that allow decision-makers to conduct more precise suppositions, finding solutions to challenges relating to climate, agriculture, water, biodiversity, quality management investigation and recommendation systems, diagnosis and treatment systems, fraud analysis and investigation, fleet management, and more. Some of the useful applications of AI are energy use by systems and meeting the demands of those systems. AI can have an impact in both cases by identifying and measuring activity at the network level and at the connected device level by sifting vast amounts of data to extract patterns and trends.

The benefits of AI applications in the use of renewal energy sources are applicable to efficiently overcome its unreliability because weather-dependent power sources such as wind and solar often fluctuate in their strength. Hence, AI can then be applied to restructure the network in real-time to minimize power consumption and provide quality, reliability, and good economy. Renewable energy can benefit as well from AI applications as it will exponentially replace fossil fuels to power ICT and AI. The energy distribution network is experiencing an increasing transformation where centralized energy systems are being decarbonized and transitioning toward distributed energy systems. Therefore, renewable energy sources along with ICT and AI will be increasingly distributed rather than centralized; energy will be produced close to where it will be used rather than at a large plant elsewhere and sent through the national grid. The benefits of AI applications can be tailored to improving power system management and ICT application's efficiency. AI applications in smart grids (SG) and renewable energy (RE) systems provide powerful tools for design, simulation, control, estimation, fault diagnostics, and fault tolerance (Bose, 2017).

Artificial intelligence has the advantage of being "flexible" to adapt, learn, and evolve in ways that conventional computers cannot because they could only operate within the parameters that were programmed into it. AI provides a competitive advantage and functionality and, with the implementation of new technologies even if they represent a major disruption in the economy, the potential to capitalize its benefits is thriving. Robotics and AI require not only an adaptation of skills but also an evolution of the daily tasks, which is essential for meeting the demands and challenges of a continuously evolving technological landscape. With the potential to accelerate SDGs, government leaders must comprehend and harness both its risks and possibilities.

Management System for Sustainable Development

A management system is a set of policies, processes, and procedures that help an organization meet the requirements expected by its stakeholders. It is based on the four-step management method Plan-Do-Check-Act (PDCA) cycle. This is an effective way to devise a strategy in any type of organization with set step-by-step guidelines for continuous improvement, bringing efficiency and optimal operability (ISO37101 Sustainable Development in Communities, 2016).

ISO 37101:2016 establishes requirements for a management system for sustainable development in communities, including cities, using a holistic approach with a view to ensuring consistency with the sustainable development policy of communities. The intended outcomes of a management system for sustainable development in communities include:

1. managing sustainability and fostering smartness and resilience in communities, while considering the territorial boundaries to which it applies;
2. improving the contribution of communities to sustainable development outcomes;
3. assessing the performance of communities in progressing toward sustainable development outcomes and the level of smartness and resilience that they have achieved;
4. fulfilling compliance obligations.

The management system for sustainable development is a useful tool that allows organizations to improve their environmental performances and to respect the sustainable development principles:

- ISO 37101:2016 is intended to help communities become more resilient, smart, and sustainable, through the implementation of strategies, programs, projects, plans, and services, and demonstrate and communicate their achievements.
- ISO 37101:2016 is intended to be implemented by an organization designated by a community to establish the organizational framework and to provide the resources necessary to support the management of environmental, economic, and social performance outcomes. A community that chooses to establish the organizational framework by itself is considered to constitute an organization as defined in ISO 37101:2016.
- ISO 37101:2016 is applicable to communities of all sizes, structures, and types, in developed or developing countries, at local, regional, or national levels, and in defined urban or rural areas, at their respective level of responsibility.

ISO 37101:2016 can be used in whole or in part to improve the management of sustainable development in communities. Claims of conformity to ISO 37101:2016, however, are not acceptable unless all its requirements are incorporated into an organization's management system for sustainable development in communities and fulfilled without exclusion (ISO37101 Sustainable Development in Communities, 2016).

Some of the benefits ISO can deliver to a city's performance:

- More effective governance and delivery of services
- International benchmarks and targets
- Local benchmarking and planning
- Informed decision-making for policy-makers and city managers
- Leverage for funding by cities
- Framework for sustainability planning
- Transparency and open data for investment attractiveness
- Comparable data for city decision-making and insight

The "smartness" of a city describes its ability to bring together all its resources, to effectively achieve the goals of quality of life. ICT has a key role in sustainable development and the sector has been a pioneer and a powerful catalyst in addressing global risks and sustainability.

Smart cities could therefore provide:

- Better and more convenient services for citizens;
- Better city governance;
- A better life environment;
- More modern industry, that is greener, and more people friendly;
- Smarter and more intelligent infrastructure; and
- A dynamic and innovative economy (ISO 37120 standard on city indicators, 2014).

Information and Communication Standardization

There will be many standardization requirements for smart cities that comprise ICTs. Some of these standards do not involve technological developments. Nevertheless, the dominant factor of ICT influence in our daily lives and how technology is rapidly influencing society as a whole means that even in domains that are not specifically ICT-centered, there are characteristics of ICTs having an impact on society, and society has an effect on them. The intertwining between the needs of attaining sustainable development, the use of new technologies in business models and infrastructure that has been driven in part by the Internet and globalization, and addressing local and global challenges such as climate change, urbanism, energy security, population growth, etc., can help the process of transformation from a city into a "smart city." A socio-economic response to petty crime, for example, can benefit from a unique approach of crime analytics, enhancing intelligence through data analysis, mapping, and real-time sharing, helping policy-makers with sufficient relevant information to take appropriate actions with the goal of reaching the desired socio-economic outcomes. Therefore, all the elements needed to transform a city into a smart city are significantly dependent on ICT (Smart Cities Preliminary Report, 2014). The technologies capable of delivering complex systems and solutions along with resilient technologies to efficiently operate within the smart city ecosystem are the essence of a sustainable smart city working toward smart solutions to environmental challenges, quality of its resident's life, and providing end-to-end solutions for the challenges ahead.

Performance and Other Indicators

The success of smart cities' strategic implementation must be defined and tracked. Thus, it is important to identify or develop sets of key performance indicators (KPIs) and other indicators to establish the criteria to evaluate ICTs' contributions to making cities smarter and more sustainable. KPIs are required to provide performance as seen from different viewpoints such as those of residents/citizens (reliability, availability, quality, and safety of services, etc.); of community and city managers (operational efficiency, resilience, scalability, security, etc.); and of the environment (climate change, biodiversity, resource efficiency, pollution, recycling rates/returns) (Smart Cities Preliminary Report, 2014). The smart city strategy of developing a road map – purpose, goals and targets, business case and impact, is also tailored to specific requirements for standardized risk assessment methodologies for critical infrastructure dependencies across organizations and sectors. Enhancing the protection and resilience of infrastructure is an urgent goal – a goal made more challenging by the inherent dependencies and interdependencies within infrastructure systems. The impression that nations' critical infrastructures are highly interconnected and mutually dependent in complex ways, both physically and through a host of ICTs (so-called "cyber-based systems"), is more than an abstract, theoretical concept, because critical infrastructures' efficient interconnections are already enhancing resilience and security in addition to facilitating the implementation of new technologies, as countries reap value from global connectivity and strive for cost-saving technologies (Harel and Baram, 2015). Dependencies and interdependencies influence all components of risk (threat/hazard, vulnerability, resilience, and consequence). They can themselves be a threat or hazard, affect the resilience and protection performance of critical infrastructure, and lead to cascading and escalating failures. In addition, critical infrastructure dependencies and interdependencies are characterized by different interactions, classes, and dimensions, making their identification and analysis both challenging and complex. Based on these factors, it is essential to integrate dependencies and interdependencies into risk and resilience assessment methodologies (Analysis of Critical Infrastructure, Dependencies and Interdependencies, 2015).

The city's leadership needs to take forward a strategic agenda allowing for the smart city governance to be measured. This process will result in maximizing efficiencies to manage, evaluate, and identify some clusters of KPIs to assess smart city governance and decision-making processes (Castelnovo et al., 2015). Before developing the road map to deliver on the smart city strategy, the city leadership should assess where the city is now, in terms of its capability to benefit from the transformational opportunities that technology can offer, and to determine the gap between this and what is needed. This includes assessing:

- the strength of the partnership structures in place and the ability of the city to coordinate the planning, implementation, and management processes effectively;
- how well key stakeholders, and particularly citizens, are engaged in helping make the city work well;
- how effectively the city is able to deliver the services that the citizen needs, including the ability to measure and monitor effectiveness;
- the quality of the connectivity assets of the city and the effectiveness of those assets in supporting collaboration and the collection and use of data; and
- how well the physical assets of the city can be used to support city management and citizen empowerment (The British Standards Institution, 2015).

Interdependencies and Dependencies Infrastructure for Sustainable Smart Cities

The facilities, systems, sites, and networks that compose a city are the solid foundation for its functional infrastructure. Therefore, the efficient delivery of essential services to the populace determine its ability to adapt and respond to risks and encompass the city's operating state and its resilience to effectively respond to disasters. Smart digital infrastructure connects data from multiple sources to generate new value and efficiencies and optimize the use of the city's limited resources. ICTs' value proposition is its ability to capture and share information in a timely- manner as well as enabling infrastructure and energy-related data acquisition, monitoring, and management operations and services. Smart infrastructure provides better decisions faster and it is the foundation for all of the key elements related to smart operations, including smart people, smart mobility, smart economy, smart living, smart governance, and smart environment. The core characteristic that underlies most of these components is that they are intertwined and generate data, which may be used intelligently to assure the best use of resources and improve performance (Smart cities and infrastructure, 2016). The term interdependencies are conceptually simple; it means the connections among agents in different infrastructures in a general system of systems. In practice, however, interdependencies among infrastructures dramatically increase the overall complexity of the "system of systems" (Gheorghe and George-Ionut, 2008) (Figure 6.2).

Smart City – Smart Leadership

Smart leadership encompasses a strategic vision and pragmatism to develop and accelerate meaningful actions for the good of the community. Local authority leaders have a key role because the community leadership role of the local authority provides them with the overarching responsibility for the way the city operates and inspires ways for collaborative efforts of all the organizations and citizens in the city to address the key priorities of the city effectively. ICTs' role is to support this collaboration, the flow of valuable information systems and digitally enhanced processes and services. In the context of diverse urban spaces, leadership that encourages ideas and innovation to bubble up from the grassroots – and where priorities on smart projects are co-determined by the full range of local agencies and residents thinking beyond the immediate project, addressing concerns for the smart city's economic, social and environmental outcomes for local businesses and residents over the long term (Smart Leadership for Smart Cities, 2016; Barrionuevo et al., 2012).

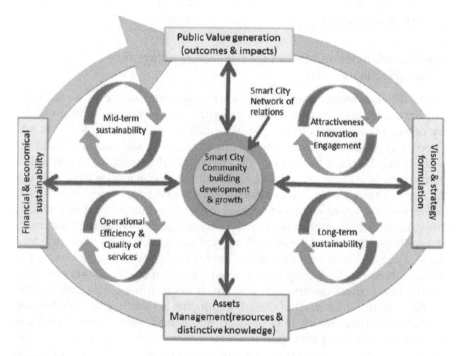

FIGURE 6.2
The logic structure of the proposed smart city's Governance Holistic Assessment Framework. (From Castelnovo et al., 2015.)

List of Terms

Artificial intelligence (AI): The study and design of intelligent agents where an intelligent agent is a system that perceives its environment and takes actions which maximize its chances of success.

Key Performance Indicator (KPI): A measurable value that demonstrates how effectively a company is achieving key business objectives. Organizations use KPIs at multiple levels to evaluate their success at reaching targets.

Public-Private Partnership (PPP): A cooperative arrangement between two or more public and private sectors, typically of a long-term nature. Governments have used such a mix of public and private endeavors throughout history.

Renewable energy (RE) (also called alternative energy): An alternative to conventional energy usually supplied by the combustion of fossil fuel such as oil, coal, or natural gas.

Smart city: A city can be defined as "smart" when investments in human and social capital and traditional (transport) and modern (ICT) infrastructure fuel sustainable economic development and a high quality of life, with a wise management of natural resources, through participatory governance (Caragliu et al., 2009).

Smart grid (SG): an electrical grid which includes a variety of operational and energy measures including smart meters, smart appliances, renewable energy resources, and energy efficient resources.

7

The Global Risk Dilemma

Digital Revolution and the Environment

Environmental performance is significant to organizations and cities to meeting stakeholder expectations on sustainability and to effectively manage the organization's energy, natural resources, and waste that have a substantial effect on its long-term success. Information and communication technology (ICT) applications are widely seen to have the potential to improve environmental performance and put organizations in a competitive advantage. The areas of the organization pertinent to manufacturing, energy, transport systems, buildings and urban systems where smart ICT applications have the potential to optimize performance and reduce inputs per unit of output. Environmental performance and green growth strategies must become part of broader economic and industrial policies, simply because economies and populations continue to grow, with accelerating global rates of production and consumption. It is necessary to innovate and develop sustainable modes of production, consumption, and living to deal with environmental challenges, and ICTs can and will play a key role in addressing these challenges (Vickery, 2012). Assessing environmental impacts is a significant step to achieving sustainability. Organizations' environmental evaluations of a product, its life cycle, can be divided into three phases: the production, the use, and the end-of-life.

At all stages, resources are extracted and emissions are inventoried to get a global picture of the environmental emissions and, therefore, the environmental impact of the product or the service. The impacts on the environment from ICT can be separated into three categories (Radermacher, 1999; Berkhout and Hertin, 2001; Fichter, 2001; Arnfalk, 2002):

- Direct effects: energy consumption, use of resources, emissions and pollution caused by the production, trade, and transport of goods, and the disposal and recycling at the end of the devices' life cycle use.

- Indirect effects: changes in the economic structure, changes of the production processes, trade, and transportation systems. The most important effects in this context are dematerialization, virtualization, and immobilization of goods.
- Effects on people's lifestyle, social values, and the "rebound effect" (Loerincik, 2006).

Sustainable Engineering for Sustainable Development

Sustainable engineering is a field of designing and/or operating systems that encourage sustainability and must be applied to entities tailoring their environmental impact assessment and procedures and should be based on principles that support sustainable development (Fedkin, 2018). The environmental, social, and economic values are the cornerstone of sustainable development and they go accordingly with the goals to attain sustainable engineering principles (Figure 7.1).

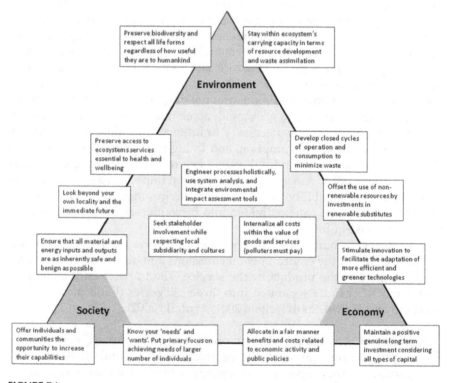

FIGURE 7.1
Classification of sustainable engineering principles versus environmental, social, and economic criteria. (From Gagnon and Leduc, 2006; Fedkin, 2018.)

Green Engineering and Sustainable Development

Sustainable development goals aim at tackling and finding solutions to global risks and new technologies, new business models, and new regulations that could potentially build a future that is different and better than the present. As the population grows, severe climatic events, resource scarcity, globalization, and environmental degradation have significantly increased the role of engineering design. It became more relevant to advance the goal of sustainability and consequently support a framework to achieve sustainable development goals. This will require a new design framework that incorporates sustainability factors as explicit performance criteria. The 12 Principles of Green Engineering developed by Anastas and Zimmerman (2003) provide a design protocol for moving toward engineering design for sustainability.

The 12 Principles of Green Engineering

Principle 1 – Designers need to strive to ensure that all material and energy inputs and outputs are as inherently nonhazardous as possible.

Principle 2 – It is better to prevent waste than to treat or clean up waste after it is formed.

Principle 3 – Separation and purification operations should be designed to minimize energy consumption and materials use.

Principle 4 – Products, processes, and systems should be designed to maximize mass, energy, space, and time efficiency.

Principle 5 – Products, processes, and systems should be "output pulled" rather than "input pushed" through the use of energy and materials.

Principle 6 – Embedded entropy and complexity must be viewed as an investment when making design choices on recycling, reusing.

Principle 7 – Targeted durability, not immortality, should be a design goal.

Principle 8 – Design for unnecessary capacity or capability (e.g., "one size fits all") solutions should be considered a design flaw.

Principle 9 – Material diversity in multicomponent products should be minimized to promote disassembly and value retention.

Principle 10 – Design of products, processes, and systems must include integration and interconnectivity with available energy and materials flows.

Principle 11 – Products, processes, and systems should be designed for performance in a commercial "afterlife."

Principle 12 – Material and energy inputs should be renewable rather than depleting (Anastas and Zimmerman, 2003).

These principles create a universal framework between designers at all scales leading to an integrated, inherent, innovative, and interdisciplinary systematic design. Organizations that develop the design, commercialization, and use of processes and products that minimize pollution, promote sustainability, and protect human health without sacrificing economic viability and efficiency can address three responsibilities – environmental, social, and economic values – cornerstones with the overarching goal of generating a balanced solution to any engineering problem.

Social, Environmental, and Financial Responsibilities – Information and Communication Technology Governance Strategy

Sustainable development must harmonize with institutions that support and facilitate a wide approach to effective, entrepreneurial, and judicious management, which can deliver the long-term success of modern society and the values it represents. In substance, governance processes refer to the quality of participation necessary "to ensure that political, social, and economic priorities are based on a broad consensus in society and that the voices of the excluded, poorest, and most vulnerable are heard in decision-making" (Governance for Sustainable Development, 2014). A strong sense of accountability must be placed in any institution to develop a solid strategy of sustainable development in the governance of ICT. Local governments must be accountable and answer to any issues and be responsible for the consequences of decisions it has made. The achievement of sustainable developments goals will raise challenges posed by social, economic, and environmental requirements in the implementation and establishment of the governance strategy (Hba et al., 2019). The implementation of an efficient strategic governance structure depends on an agile model and horizontal managerial culture. Moreover, institutions and organizations adopting Corporate Sustainability and Responsibility (CSR) and a comprehensive holistic inclusion of Green information technology (IT) practices will lead to the creation of a more sustainable environment. This approach supports the development of the green governance as a new generation model of management and best practices. Leaders need to emphasize the commitment of actions toward a legal and regulatory framework. If decisions and policies are not anchored in a binding and enforceable framework, they become merely ornamental in nature (Qalyoubi, 2012).

Institutional structures are conditioned by many socio, environmental, and economic elements influencing our daily lives. These variations in institutional structures and the different cultures influencing management styles and all technological and socio-political considerations inherent to the field of ICT leads to wide-ranging differences in ICT governance. The University of Nottingham, UK, published a research highlighting the practices that have been found to improve the delivery of strategic ICT. The following shows a summary of their findings across four areas:

Institutional Synergy

Fragmented institutional structure presents challenges for governance and the growing importance of ICT in supporting institutional strategy that aligns policies, accountability, and mechanisms that allow each member of the organization to participate in the desired objectives is fundamental to provide agility to adapt and lead with transparency and accountability. Therefore, taking advantage of changes as opportunities and rapid adaptation in response to new challenges is a priority to good governance as people will have more confidence in local governments that act in the community's overall interest, regardless of differing opinions.

Institutions exhibiting well-aligned ICT strategy displayed the following actions:

1. Assess the current investment strategy and understand why any new investments have not been previously identified within the strategy.
2. The formulation of a documented and approved institution-wide ICT strategy. Some institutions are formulating separate IT and Institutional Synergy (IS) strategies.
3. The cross-reference of the ICT strategy to reinforce alignment to the institutional strategy
4. Using a process for review and updating of the strategy
5. The horizon for institutional planning is changing and the ICT strategy needs to reflect this as a working document. To become more flexible and agile it must synchronize with institutional strategic planning and change.
6. The synergy required for the alignment of ICT and institutional strategies across strategic and operational planning and reporting is concisely illustrated by the following example of an institution's strategic alignment map.

Governance Decisions and Mechanisms

The efficiency of governance is essential for any institution if it wants to set and meet its strategic goals. The governance structure combining controls, policies, and guidelines that drive the organization toward its objectives while also satisfying stakeholders' needs is often a combination of various mechanisms. ICT governance is the system by which the current and future use of ICT is directed and controlled; therefore, it needs the coordination of the institution-wide ICT governance structure to accommodate an institution's management and ICT delivery structures.

Some of the highlighted actions for good ICT governance are exhibited in the following:

- The ICT strategy was approved and implemented by a governing committee made up of senior management, Chief Information Officer or an equivalent role, together with representative management from across the institution. This supports institution-wide awareness and buy-in. We found this committee to exist under various differing titles and we shall refer to it as the Information Systems Strategy Group (ISSG).

- Decision-making in relation to portfolio management was improved by introducing the use of a Portfolio Decision-Making Matrix that identified the appropriate positioning of services and solutions as "bleeding edge," "leading edge," "on a par," or "lagging behind."

- ICT strategy formulation was conducted with institutional-wide consultation involving senior managers and unit heads.

- The ICT strategy was approved by the Senior Management Team.

- All ICT investment was approved by the ISSG.

- Capital and operational budgets were allocated and approved by the ISSG.

- An exception process was established (development without architecture). This ensured that exceptions to strategy were visible and, if agreed, proceeded against an approved business case that provided greater understanding and feedback on the value and cost of exceptions.

- ICT principles, policies, and standards defined and adhered to. It is expected that these will facilitate better investment proposals, progress reporting and measurements for value and ROI (return on investment) and therefore support improved accuracy and availability of information to assist decision-making and management.

Governance Communications

An effective communication is vital to the successful delivery of ICT projects. However, the ability to address and balance the priorities within strategic planning highlights the need for such communications between institutional management. In addition to consultation on strategic requirements, there are other techniques that have been found to enhance institutional awareness and buy-in to strategy:

- Obtain senior management team buy-in and promotion of ICT governance

- Use committees across the institution to add awareness and create influence
- Use a Chief Information Officer and a compliance function to own and promote ICT governance
- Identify and try to win over management who don't comply
- Provide a portal or other home for ICT governance information to ease its use and assist in visibility

Governance Performance

Governance allows for the measurement of performance in two areas:

- Services performance.
- Definitions for ICT service levels and project progress reporting provide both project and operational management and reporting to the ISSG and senior management team. Service levels defined and agreed as part of the ICT governance were found to be actively used for service communications and monitoring. However, interviewees commented that more work was required to establish improved metrics and reporting in relation to project progress and final delivery against objectives.

Performance against Institutional Strategy

This is a quick assessment of the senior management team to gain a measure of how well ICT governance is delivering ICT services that meet the core institutional strategic objectives. Each senior manager will have his or her own perspective which will inform against an overall assessment.

The assessment typically requires:

1. The definition of a set of strategic objectives or outcomes, for example, cost-effectiveness, transformation, business improvement, or agility
2. Each member of the management team to assess for their domain:
   a. the importance of each of the outcomes;
   b. the influence of governance on the success of each of the outcomes;
   c. where and why is governance effective;
   d. where and why is governance less effective.

The information provides an additional visible measure of performance as a basis for senior management discussion and review of governance. Although a frequent change in ICT governance is not desirable, governance is specific to institutional structure and management and should be regularly reviewed to ensure that it continues to meet an institution's unique requirements (Strategic ICT, 2012). There are several overlaps between sustainable development and good governance. Sustainability is not just the conservation of natural resources and minimizing environmental impacts. It is about meaningful advancement. Nelson (1999) stated that technological change has been the central driving force behind the economic growth we have experienced; this is still relevant today. Nevertheless, technological advancements in a society build a sense of experiencing "perpetual economic growth"; that is, the notion of economic growth bringing equality and prosperity into society. However, sustainability, and the goal to achieve sustainable development, implies that the benefits of growth are more than just economic. Social development and protection of the environment are also fundamental elements for acquiring economic justice. The requirements to achieve sustainable development are synergistic. The core pillars consist of maintaining the integrity of biophysical systems; offering accessibility to basic services; and deterring hunger, nuisance, and deprivation to people (Kemp et al., 2007; Kemp et al., 2005).

List of Terms

Corporate social responsibility (CSR): A business approach that contributes to sustainable development by delivering economic, social, and environmental benefits for all stakeholders. CSR is a concept with many definitions and practices.

Green engineering: The design, commercialization, and use of processes and products that minimize pollution, promote sustainability, and protect human health without sacrificing economic viability and efficiency.

Sustainable engineering: The process of designing or operating systems such that they use energy and resources sustainably, in other words, at a rate that does not compromise the natural environment or the ability of future generations to meet their own needs.

8

ICT Strategic Framework

Strategic Approach of Information and Communication Technology to Cope with Global Risks and Advance Economic and Social Progress

Addressing risks are an increasingly important part of management for the public and private sector because they include such threats as network, telecommunication and transportation interruption, disaster recovery, severe weather events impacts and discontinuity of systems, business and information security threats, data backup and protection, and public health risks and more. It is fundamental to create the right incentives and encourage actions by individuals that reduce collective risks. In that context, creating social, environmental and economic incentives to conserve biodiversity and develop concrete actions toward economic growth and social progress is of paramount importance because we live in a world that economic growth does not take proper account of environmental effects. Capitalizing in strong institutions and coordinated international action show a path to sustainable development goals (SDGs), which in essence needs concerted international efforts, with adequate representation which are required to accelerate world production growth, to facilitate the flow of goods and services and to support effective resource utilization. The World Economic and Social Survey (2017) emphasized important factors that contribute to develop an effective strategic approach to sustainable development and advance economic and social progress. The following three factors are significantly relevant to guiding countries through a difficult current global economic situation and for the implementation of the 2030 Agenda for Sustainable Development:

- Stability in the international monetary and trade systems underpins development. In that respect, the Survey has consistently highlighted risks associated with volatile commodity prices and warned against protectionism. Regarding the international monetary system, the Survey has advocated for a shift away from a single-currency system and called for effective financial regulation and supervision.

- Countries need adequate policy space to accelerate development. Flexible application of the international norms and commitments has facilitated economic recovery at times of crisis and major adjustment. In the late 1950s, the flexibility shown by the International Monetary Fund (IMF) toward Western European countries was a determinant in these countries' success in eliminating foreign exchange restrictions and establishing current account convertibility.

- Development is multidimensional, context-specific and about transformation, underpinned by strategic development planning and strengthened State capacity. Proper coordination across various policy areas and diverse actors are needed in bringing about structural and institutional changes, which would lead countries toward economic diversification, stable growth and improved living standards (World Economic and Social Survey, 2017).

The risks to attain sustainable development goals are greater than ever as the world witness nations' progress that has been uneven and insufficient. Environmental sustainability is under threat, with accelerating growth in global greenhouse gas emissions and biodiversity loss. Furthermore, one of the SDGs is under threat and face difficult challenges as more than a billion people still live in extreme poverty, putting the agenda of eradicating poverty on the front page of the central focus to the international development program. It is necessary to implement pragmatic actions in a transparent and efficient way and make equitable and balanced economic growth and human welfare in order to preserve economic, social, and environment foundations where the present and future generations can work closely together, building mutual trust and the distribution of responsibilities among all the stakeholders for a common goal. That translates into setting priorities and what constitutes an appropriate number of goals with a process system that entails monitoring, reviewing, and adjusting procedures, with continual improvement, and creating values that make a harmonious society among all actors. Public and private sectors must be involved in designing resource management planning processes. This is an area of strategic planning and risk management that is aligned with the overall goal of institutions of all areas to efficiently achieve SDGs. It has a direct relationship in the process of developing and implementing the goals and it is an ongoing activity throughout the planning cycle to assess how well the plan is working. Countries are facing a growing number of challenges to achieve SDGs and risk management strategies are highly relevant to those goals. Developing options and actions to enhance opportunities and reduce threats to desired objectives tailored policy and plan effectiveness monitoring and helps determine the need for future actions to achieve efficiency, and possible changes and improvements in policy and decision-making and planning. Hence, building a strategic approach to risk management support dealing with diverse risks, which is needed for accelerated poverty reduction, and sustained economic and social development must be an objective tailored to policy and planning. As it is increasingly evident a model of human development based on economic progress alone is not sufficient to have a sustainable society, the social factors affecting populations needs to be addressed with tangible solutions. The aforementioned risks and its solutions within an economic and social context are showing a systematic series of actions with a common denominator: we have entered a dangerous new phase of the global crisis. Without collective resolve, the vision and solutions that the world so badly needs to materialize actionable strategies to cope with risks will not be capitalized. The risk of core instabilities in developed and developing countries is still present. We live in an interdependent world; economic tremors in one country can resound swiftly and strongly across the globe, particularly if they originate in systemic economies that consequently affect many areas of society. To be sustainable, development must be economically sustainable (or efficient), socially desirable (or inclusive) and ecologically prudent (or balanced) (Romeiro, 2012). Therefore, the risk of "core instability" deters the working progress of SDGs as financial instability and social and environmental risks have risen significantly. The United Nations Research Institute for Social Development published a paper in 2014 that identifies the social drivers of sustainable development.

- Social policy as a driver of transformation;
- Social reproduction and the economy of care;
- Equality of opportunity and of outcomes;
- Employment-centered economic policy;
- Social and solidarity economy;
- Empowerment, effective participation, and accountability (Emerging Issues: The Social Drivers of Sustainable Development, 2014).

Information and Communication Technologies (ICTs) can facilitate and support a broad-based social and economic development. Hence, ICTs can be used as a tool to achieve a variety of development objectives, including poverty reduction, extension of health services, expansion of education opportunities, and access to government services. ICTs are a fundamental asset to economic development on the national and international level, by improving developing countries' positions in the global economy and raising individual levels of income. It has become the foundation of every sector of every economy, everywhere (William et al., 2007).

Some of the roles of ICTs in modern economic growth and development are:

- reduce transaction costs and thereby improve productivity;
- offer immediate connectivity – voice, data, visual – improving efficiency, transparency, and accuracy;

- substitute for other, more expensive means of communicating and transacting, such as physical travel;
- increase choice in the marketplace and provide access to otherwise unavailable goods and services;
- widen the geographic scope of potential markets; and
- channel knowledge and information of all kinds.

Some significant benefits have been researched and disclosed at the macro-level growth coming from information technologies, telecommunications and mobile communication investments in developed and developing countries. At the level of the firm, World Bank surveys of approximately 50 developing countries suggest that "firms using ICT see faster sales growth, higher productivity and faster employment growth" (Khalil et al., 2008).

Another reason of the attributes of ICTs in furthering the sustainable development agenda is its value and impact that can be adapted to other economic and social sectors. Three ICT capabilities are especially important for economic and social development:

1. Enabling greater efficiency in economic and social processes;
2. Enhancing the effectiveness of cooperation between different stakeholders;
3. Increasing the volume and range of information available to people, businesses and governments (Information and Communications Technologies for Inclusive Social and Economic Development, 2014).

Partnerships are key to implement a sustainable development agenda and the full range of stakeholders in international development involvement is needed to effectively implement a strategy that encompasses governments, both industrialized and developing, the business and non-profit sectors, multilateral agencies, and community organizations on the ground (Sustainable Development Knowledge Platform, 2015). The implementation of strategies with holistic, coherent, and integrated approaches at the national, regional, and global levels with corresponding policies addressing interlinkages between the social, economic, and environmental dimensions of sustainable development are not fully developed and implemented. Strategy implementations require many steps with detail action plans and specific work assignments with dedicated personnel and process, assessing human capabilities, monitoring actions and continual improvements to find the desired outcomes. Financial investment is central in order to have the program running successfully, first by calculating estimated figures, then by refining those into more detailed costing commitments that can be expressed in budgets and business plans.

Strategy and Information and Communication Technology Frameworks to Tackle Global Risks

The new, dynamic, and fast-growing ICTs pose a new organization experience. Under business as usual models, the investment in ICTs is uncoordinated across the enterprise and leaders needs to implement a coordinated, strategic and focus-oriented plan without the typical fragmentation and inconsistency witnessed across governments and private entities. The recognition of weaknesses across the public and private sector is key to develop an enterprise-wide information strategy.

Information and Communication Fragmentation

The ability to deliver effective communication and clearly develop a strategy without losing coherence to implement your plan relies on the capacity of nations and private organizations to capitalize on the benefits of delivering substantive efforts to achieve sustainability. I refer to Information and Communication Fragmentation (ICF) as the significant challenges *impeding* technology adoption and lack of resources and tools to efficiently deliver your strategy and consequently can lead to cognitive overload. That is too much information and too many tasks that need to be accomplished it can result in the inefficient use of time and focus that exacerbates risks and also stress and other social problems impending the overall outcome. ICF also refers to the lack of knowledge about the issues you are confronting and can lead to a lack of overview. This issue can be counteracted by technology. The aforementioned issues affect the overall strategy to achieve nation's goals in sustainable development and show us that strategies and plans for integrated and holistic implementation of 2030 Agenda for Sustainable Development and the SDGs happen best under conditions that are aligned with human cognitive architecture. There are many limitations in the ICT sector to support a system to achieve SDGs. The narrow scope and weaknesses in existing information systems, a multiplicity of data collection systems designed, and the lack of capacity for data analysis are among the limitations to be addressed (Nabyonga-Orem, 2017), particularly in developing nations. Therefore, strong leadership and a comprehensive and longer-term approach to strengthening the existing strategies and develop new ones is key to address and overcome the existing ICTs limitations and obstacles to achieve the aforesaid goals. Sustainable development inherently involves many contrasting stakeholders from different sectors of society such as governments, private organizations, nongovernmental organizations (NGOs), etc., and strengthening country capacity to monitor SDGs will involve several actions that can help the success of the goals' outcome. Access to advanced technologies has grown at a fast pace; however; there are existing gaps in ICT access – between and within countries, between urban and rural settings, and pose a major digital divide particularly poor access in vulnerable developing nations. As information communication fragmentation is still an issue that is not fully addressed with the addition of the digital divide the challenge now is to develop and implement policies tailored by strategies to diminish the gap, providing tools and resources with "a domestication of the SDG agenda through country-level planning and monitoring frameworks, prioritizing interventions, indicators and setting country-specific targets" (Nabyonga-Orem, 2017).

Designing an Information and Communication Technology Strategy and Its Implication to Sustainable Development

The Sustainable Development Goals (SDGs) represent a renewed commitment on the part of leaders around the world to deliver substantial changes to the decision-making process and policy. Your strategy is the road map to successfully implement your vision and goals and how ICT can support and accelerate progress toward SDGs. Therefore, it is important to design an ICT strategy fit-for-purpose needs to assess the current ICT capability, looking at emerging trends and identifying opportunities in emerging technologies. To realize the potential that ICT solutions offer in understanding progress toward SDGs, it is necessary to develop a "data culture" that put emphasis in data visualization and analytics process with a commitment to use those insights to make changes in the way the organization works and interacts (Sustainable Development Goals ICT Playbook, 2015). The major constraint to achieve SDGs is a lack of good data and strategic planning to implement the crucial changes to make a pragmatic impact. Very few of the proposed SDGs indicators have good data and almost none have thresholds. Moreover, there are not enough quality studies on the impact of ICT to support neither SDGs nor initiatives that are adapted to each country's strengths and vulnerability to efficiently deliver the aforementioned goals. Therefore, it is important to composing the strategy around the three sustainability pillars: Economic, Social and Environmental schemes because it can help to deliver the initiatives and gather and share better ICT data.

Strategic Sustainable Development—Synergies of Applied Tools

There are important steps to take to find the appropriate strategic approach to deliver a designed-management tool to strategize sustainable development goals applications:

- Assess ICT operations and current state to find evidence of the impact of ICT solutions in place.
- Identify ICT solutions on specific SDG goals, targets, and especially indicators.

- Develop relationships and collaborations between the sector to establish standards for mapping solutions.
- Establish an impact assessment and key delivery principals.
- Tailor to SDCs the vision and culture of the organization aligned to sustainability principals and desired outcome.

Frameworks and guidelines have been used in management for decades to improve operations and designed strategic models for efficiency. Tools for management and monitoring of sustainable development have gained worldwide acceptance in the last decade, like International Organization of Standardization (ISO) 14001, Life Cycle Assessment (LCA), ecological foot-printing, Factor 4, Factor 10, Sustainable Technology Development, Natural Capitalism, and The Natural Step Framework. They have been supported by a number of organizations and programs (Robèrt et al., 2002). These tools for management and monitoring provide clarity to develop the strategy and improvements to efficiently advance the agenda.

Information and Communication Technology and the Transition toward a Circular Economy and Sustainable Development

The exponential growth of populations, scarcity of natural resources, environmental degradation, and climatic events put an alarming emphasis on sustainable development strategies to tackle the global risks the world is confronting. The efforts to achieve solutions come from local and national governments, private organizations, and international cooperation, and is the crucial role of the decision-making leaders. The circular economy is a new economic model balancing economic development with environmental and resources protection (Jawahir and Bradley, 2016). In substance, "A circular economy is restorative and regenerative by design and aims to keep products, components, and materials at their highest utility and value at all times. The concept distinguishes between technical and biological cycles" (Kriza, 2017). It promises pragmatic solutions to cope with the aforementioned risks in a smart and innovative conceptual approach, the efficient use and reuse of resources and a strong business case. The strategy in place is the foundation of achieving SDGs through embedding the 3R principles of materials use – reduce, reuse and recycle – into production and consumption process, the circular economy can accomplish greater resource productivity, higher energy savings, and lower greenhouse gas (GHG) emissions (Vasiljevic-Shikaleska et al., 2017). This creates greater value by aligning incentives through a change in business models and identifying the management tools that build on the interaction between products and services by incorporating digital technology to track and optimize resource use. It also strengthens the connections between supply chain actors by using digital, online platforms and technologies that provide an understanding of how to strategically proceed to accomplish the goals. Societies must have economic growth within an effective balance between economic developments, environment, and natural resource preservation. The aforementioned challenges and environmental burdens are sending many countries to seek innovative approaches to address these problems as a means for implementing appropriate policies.

Blockchain Technology and Sustainable Development

The Fourth Industrial Revolution is happening now with rapid advances in technology and artificial intelligence, and innovations are becoming faster, more efficient, and more widely accessible to the general population. Moreover, technology is also becoming increasingly connected; in particular, we are seeing a merging of digital, physical, and biological realms. New technologies are enabling societal shifts because of its effect on socio-economic developments, values, identities, and possibilities for future generations (The Fourth Industrial Revolution for the Earth, 2018). Although the world is experiencing great progress in information technology we are still on the fringes of this revolution. The question is how the fourth revolution will support, influence, and build a sustainable world and change the status quo for inclusive growth and sustainable development.

Blockchain is a rapidly evolving area of information technology, with the potential for huge benefits in terms of security, reliability, and cost-efficiency in the exchange of information (Exploring the transformative potential of blockchain for sustainable development, 2017). Blockchains also have unlimited potential to disrupt industrial sectors, commercial processes, governmental structures, or economic systems. Many governments and private institutions perceive the transformative power of blockchain technology as a threat to existing systems of governance. To the contrary, it is an opportunity for national and international institutions to defend the rights of those they represent and to accelerate the collective progress toward meeting the United Nations' Sustainable Development Goals (The Future is Decentralised, 2018). To achieve the SDGs there is a need for accountability, responsibility, and transparency in government systems, and verifiability and immutability for commercial processes. Blockchain is able to shape the current system and bring novel applications of cryptography and information technology to age-old problems of financial record-keeping, leading to far-reaching changes in corporate governance (Yermack, 2017). Sustainable development requires a strategic approach to solve many issues related to the resilience of infrastructures, guarding privacy and at the same time guaranteeing autonomy, and encourage cooperation in innovation and rethink the information system.

Basden and Cottrell (2017) stated the benefits of blockchains, highlighting how utilities are using blockchain to modernize the grid, providing a reliable, low-cost way for financial or operational transactions to be recorded and validated across a distributed network with no central point of authority. In brief, blockchains can help further the agenda to build economic, environmental, and humanitarian initiatives in pursuit of the SDGs.

Managing Risks Associated with Information and Communication Technology

The infrastructure and capacity of information technology to deliver safe, effective, timely, efficient, and equitable services is becoming an increasingly essential part of business operations. Hence, cloud computing models are emerging as the solution to all the infrastructure setup problems of IT industry and green cloud computing as a source of creating value and reducing power consumption by using virtualized computational resources to provide an application's computational resources on demand. Auto-scaling is an important cloud-computing technique that dynamically allocates computational resources to different applications. It works well for facilitating additional capacity, but they do not address application performance issues. Some of the benefits are the capacity of matching their current loads accurately with actions that lead to the removal of resources that would otherwise remain idle and waste power (Dougherty et al., 2012)

Energy consumption and efficient energy delivery are central to nearly every major challenge and opportunity as it transforms lives, economies, and our environment. In regard to cloud infrastructure, its demand is increasing exponentially because of the energy consumption of data centers. When the energy consumption of the data center increases, carbon emissions rise and detrimentally impact the environment and our health, just to mention a few. Worldwide spending on public cloud services and infrastructure is forecast to reach US$160 billion in 2018, an increase of 23.2% over 2017, according to the latest update to the International Data Corporation (International Data Corporation, 2017). Although annual spending growth is expected to slow down, the market is forecast to achieve a five-year compound annual growth rate (CAGR) of 21.9% with public cloud services spending totaling $277 billion in 2021 (International Data Corporation, 2017).

Cyber-Security and Risk Management: An Imperative Agenda for Sustainable Development

Our daily life, economic and social strength, human health, the environment and national security depend on a stable, safe, and resilient cyber-space. Sustainable development is not foreign to the risks of cyber-attacks and it calls for urgency in building cyber-security awareness and a comprehensive strategy to cope with those risks. There is a link between cyber-threats and climatic events. The US security establishment increasingly understands climate change and its direct and indirect impacts on water, energy, and infrastructure. Sea lane security and cyber-threats in the form of cyber-warfare and cybercrime are emerging transnational threats that must be taken just as seriously as other threats (Allen, 2014).

As cyber-attacks are emerging threats that have the potential to cause business disruptions, financial chaos, waste discharges, catastrophic spills, and air emissions, just to mention a few. Information technology security strategy and technology risk management frameworks provide directions to build the appropriate strategy while adopting advanced technologies. The use of The National Institute of Standards and Technology (NIST) guidelines, ISO 31000, ISO 27001, and ISO 22301 frameworks are becoming the strategic tools of choice to evaluate risk, analyze the likelihoods and consequences of risks, secure investments, and develop systems' continual improvement for the most pressing security initiatives.

A culture of cyber-security helps to build trust in the digital environment and support the SDGs. Hence, risk management frameworks provide a wide-ranging assessment and of the current situation and determine their current cyber-security capabilities, shaping a policy framework of computer security guidance for how private and public entities can improve their competences to cope with cyber-vulnerabilities and establish a cyber-security strategy. Moreover, frameworks help understand the maturity of security activities and can adapt over time to meet the maturity level of the threats faced and the security capabilities employed. Risk management assessment based on the valuation of cyber-risks must be able to deal with different types of uncertainty to achieve security and sustainable development goals without jeopardizing the strategic planning in place, the international standards must be an essential part of the strategic objectives.

National Institute of Standards and Technology Cyber-Security Framework

The National Institute of Standards and Technology's Cyber-Security Framework (NIST CSF) is a voluntary framework that consists of standards, guidelines, and best practices to manage cyber-security-related risk. In essence, the cyber-security framework's prioritized, flexible, make risk decisions, take action to reduce risk and cost-effective approach and support concrete steps to protect critical infrastructure and build resilience (NIST cyber-security framework, 2018).

National and economic security must be regarded and protected to achieve SDGs, as the development and implementation of frameworks to safeguard the well-being of nations are vital to achieving sustainable development. The following figure developed by NIST shows the core functions of the framework that can be adopted by organizations across the United States, as well as internationally. It is significant for achieving security to have a step-by-step process of continual improvement (see PDCA cycle) to better understand and minimize risks. The five functions included in the NIST Framework Core are:

- Identify
- Protect
- Detect
- Respond
- Recover

The functions are the highest level of abstraction included in the Framework Core. They act as the backbone of the Framework Core that all other elements are organized around. These five functions were selected because they represent the five primary pillars for a successful and holistic cyber-security program. They aid organizations in easily expressing their management of cyber-security risk at a high level and enabling risk management decisions (Figure 8.1).

The five Framework Core functions:

- *Identify*: Develop the organizational understanding to manage cyber-security risk to systems, assets, data, and capabilities. The activities in the Identify function are foundational for effective use of the Framework. Understanding the business context, the resources that support critical functions and the related cyber-security risks enables an organization to focus and prioritize its efforts, consistent

FIGURE 8.1
Cyber-security framework version 1.1. (*Source*: NIST, 2018.)

with its risk management strategy and business needs. Examples of outcome categories within this function include Asset Management, Business Environment, Governance, Risk Assessment, and Risk Management Strategy.

- *Protect*: Develop and implement the appropriate safeguards to ensure delivery of critical infrastructure services. The Protect function supports the ability to limit or contain the impact of a potential cyber-security event. Examples of outcome categories within this function include Access Control, Awareness and Training, Data Security, Information Protection Processes and Procedures, Maintenance, and Protective Technology.
- *Detect*: Develop and implement the appropriate activities to identify the occurrence of a cyber-security event. The Detect function enables timely discovery of cyber-security events. Examples of outcome categories within this function include Anomalies and Events, Security Continuous Monitoring, and Detection Processes.
- *Respond*: Develop and implement the appropriate activities to act on regarding a detected cyber-security event (National Institute of Standards and Technology, 2018).

It is important to emphasize that standards need to be neutral in order of limiting undesirable effects. *Technology neutrality* can be used in connection with standards designed to limit negative externalities (Maxwell and Bourreau, 2014). Therefore, this framework remains effective and supports technological innovation because it is technology neutral. The use of existing and emerging standards will enable economies of scale and drive the development of effective products, services, and practices that meet identified market needs. It is important to notate the significance of market competition leading to the promotion and development of new and existing technologies and practices, and the realization of significant benefits by the stakeholders. Building from those standards, guidelines, and practices, the framework provides a common categorization of actionable steps and mechanism for organizations to:

1. Describe their current cyber-security posture;
2. Describe their target state for cyber-security;
3. Identify and prioritize opportunities for improvement within the context of a continuous and repeatable process;
4. Assess progress toward the target state;
5. Communicate with internal and external stakeholders about cyber-security risk (Cybersecurity Framework, 2018).

Nations and organizations addressing cyber-security need to build investigative techniques and legal tools together in combating these threats.

Nevertheless, as a key part of its systematic process for identifying, assessing, and managing cyber-security risks, they also need to develop frameworks that complement each other considering the global risks implications of inaction. Cyber-security risk management will enhance critical service delivery and prioritize expenditures to maximize the impact of the investment and better manage their cyber-security risks (Cybersecurity Framework, 2018).

ISO 31000 and ISO 22301: Managing Risks in the Digital World

While organizations have been conducting risk assessments for years, many still find it challenging to obtain their real value. A strong business case that applies risk management ISO 31000 and the business continuity management system standard (ISO 22301), and recommends a systems view of risk assessment and proactive approach to risk management through a shared response at local and international levels, would become increasingly important to measuring cyber-security risks. It is necessary to establish the context and set of risk management assessment objectives (Marolla, 2017). To set the objectives of a risk management assessment, the location and scope of the study and the operating processes in the area under threat have to be established. An assessment seeks to identify and investigate risks to describe the actions that are required to attain the following objectives:

1. The potential impact on
2. A particular value from
3. A threatening process (Rollason et al., 2011).

Risk criteria are also important considerations for the framework. Likelihoods and consequence scales and their combination in the current conditions must be included when defining the acceptable risk level. Stakeholders should determine what that level is and then identify tolerable and intolerable risks (giving priority to the latter) that need to be addressed according to local conditions (Rollason, 2011).

Risk-modeling techniques are increasingly used by many governmental entities to evaluate exposure to cyber-attacks. Assessing and comparing different types of risks from adverse impacts becomes fundamental in understanding and minimizing their impacts and recover quickly after the event's occurrence on the balance of probabilities.

There are three central risk-modeling technique inputs:

- *Hazards*: Cyber-attacks impacting an organization or massive cyber-attacks on nations can present uncertainties about its hazard occurrences and the affecting systems and facilities exposure. Existing knowledge of past events on a local or global context affords a tangible concept of the intensity and frequency of what is expected.

- *Exposure*: Location and geographical distribution of the territory affected by a hazard where the organization is located need to be mapped. An account of the current capabilities, human occupation, and physical assets has to be performed. Particular components (e.g., protect valuable data, monitor upcoming cyber-risks, understanding your "cyber-perimeter") must be identified to differentiate the exposure.
- *Vulnerability*: Exposed elements/systems susceptibility is directly linked to the level of hazard. Cyber-attacks can be differentiated and analyzed by the intensity and the frequency of the impacts.

Cyber-threat models are frequently little more than a progression of semi-descriptive labels: hackers, hacktivists, script kiddies, nation-states, cyber-terrorists, organized crime, or malicious insiders. Hazard parameters can be established by analyzing and understanding past records of events and risk modeling probabilities. The nature, magnitude, frequency, and intensity can help determine the level of the hazard; consequently, hazard models, which are based on a set of assumptions that should be conveyed to the model user, may present a reasonable account of complex dynamics and evolution of the threats (Improving the Assessment of Disaster Risks, 2012; Marolla, 2017).

Principles of Risk Management: The Process of Creating Value

Creating a culture where the city's workforce addresses risk in every activity, is crucial for effective plan development, implementation, and preparedness for any event. It builds common terms and metrics for addressing risks, increasing the chances of efficient actions toward climate change impacts on their population without adopting an unwarranted risk-averse position. The 11 principles of risk management, as defined in ISO 31000, need to be considered in this process. This standard helps to accomplish the following goals:

1. Create and protect value.
2. Assure risk management is an integral part of all organizational processes.
3. Shape decision-making at all levels in the organization.
4. Manage uncertainty.
5. Ensure systematic, structured, and timely responses.
6. Take advantage of the best available information.
7. Make sure risk management is tailored to each organization.
8. Account for human and cultural factors.
9. Assure transparency and inclusiveness.

10. Provide a dynamic, iterative and responsive to change.
11. Facilitate continual improvement of the organization.

Risk management frameworks, and specifically ISO 31000, can be highly useful in identifying and treating the aforementioned risks. One specific major issue with cyber-attacks is trying to distinguish between the current situation and the pragmatic approach of what we can actually achieve. The first step is to critically examine the objectives that are being set. Are they in fact achievable, or are they a desirable pie in the sky hope? Will they actually provide meaningful outcomes, or are they essentially feel-good images? This is where ISO 31000 applies the process for testing the objectives and developing an effective understanding of what uncertainty exists and how or if it can be controlled within the resources and knowledge available to the program.

ISO 31000 provides the information needed to establish the principles, develops the right changes to the management system to be employed, and provides a process for understanding and managing the risks that will imperil the achievement of our objectives with respect to protecting the population from aspects of climate change. It also helps to pinpoint the factors we can change and those we cannot, as well as to know the difference between the two (Marolla, 2017; Knight, 2013).

Different Stages of Risk

Risk management can identify the different stages of risk and develop measurement processes for climate cyber-security. The framework to identify risk is recursive and can be evaluated periodically. The following steps are important in selecting risk management processes to evaluate cyber-security vulnerabilities, and develop strategies to cope with risks for real-time response:

1. A scoping exercise where the context of the assessment is established. This identifies the overall method to be used.
2. Risk identification. This step also identifies scenario-development needs.
3. Risk analysis, where the consequences and their likelihood are analyzed. This is a well-developed discipline with many methods available to undertake impact analysis.
4. Risk evaluation, where contingency methods are prioritized.
5. Risk management or treatment, where selected cyber-security measures are applied.
6. Monitoring and review, where measures are assessed and the decision made to reinforce, re-evaluate, or repeat the risk assessment process (Jones and Preston, 2010; Marolla, 2017).

Analyzing Risk

The assessment framework is developed to methodically analyze the risks. Then, the risk is evaluated by identifying the consequences and probabilities in the context of existing controls. Considering the source of risk is important and it has to be emphasized to understand the positive and negative effects and probabilities. An individual analysis of the consequences and likelihoods is needed to perform a qualitative climate-change risk assessment based on particular outcomes (Marolla, 2017).

The risk analysis stages are as follows:

1. Analyze existing controls – Identifying existing controls to minimize the consequences and likelihood of each risk. Only existing controls that are funded and completed should be measured in this stage.
2. Analyze the event's magnitude of consequence and likelihood – Determine the phase of the magnitude of an event's consequence and its likelihood of occurring.
3. Assign the risk priority rating – Using the risk priority shown in Table 8.1, the risk rating can be obtained. The process of analyzing different scenarios has to be put into practice for each risk (Marolla, 2017).

ISO 31000 identifies the level of risk, expressed in terms of the likelihood of an event and its consequence. A description of the likelihoods and consequences to define the level of risk is fundamental to developing a risk management strategy (Rollason, 2010).

ISO 31000 classifies risk evaluation as a framework to compare the results of the risk analysis with risk criteria and to determine if the level of risk is acceptable, allowable, or intolerable. The priority is given to intolerable risks. It is impossible to treat every risk and there is a possibility that high implementation costs might offset the benefits or risk reduction achieved. The methods of reducing risks are evaluated and the actions to investigate new management measures are put in place. The integration of the likelihood and consequence in the previous step presents the "unmitigated risk": risks that are not diminished or moderated in intensity or severity. After implementing existing management measures in the assessment, identifying risk priorities that need immediate attention to take place (Rollason, 2010).

Treating the Risk

Risk treatment involves developing a range of options for mitigating the risk, assessing those options, and then preparing and implementing action plans.

TABLE 8.1

Risk Priority Ratings Example (Given that a Scenario Arises)

Risk Assessment Matrix		Probability (Expected Frequency)				
Severity (expected consequence)		Frequent: Continuous, Regular, or Inevitable Occurrences	Likely: Several or Numerous Occurrences	Occasional: Sporadic or Intermittent Occurrences	Seldom: Infrequent Occurrences	Unlikely: Possible Occurrences but Improbable
		A	B	C	D	E
Catastrophic: Mission failure, unit readiness eliminated; death, unacceptable loss or damage	I	EH	EH	H	H	M
Critical: Significantly degraded unit readiness or mission capability; severe injury, illness, loss or damage	II	EH	H	H	M	L
Moderate: Somewhat degraded unit readiness or mission capability; minor injury, illness, loss, or damage	III	H	M	M	L	L
Negligible: Little or no impact to unit readiness or mission capability; minimal injury, loss, or damage	IV	M	L	L	L	L

Legend: **EH**, Extremely High Risk; **H**, High Risk; **M**, Medium Risk; **L**, Low Risk.
Source: Adapted from U.S. Army Composite Risk Management (CRM), 2017.

These treatments can follow a series of principles that will positively and efficiently impact the implementation of the framework:

- Keep a balance between different levels of risks.
- Focus on high priority risks to support the strategic planning.
- Implement small, flexible, and incremental changes based on regular monitoring and revision of plans.
- Keep options open for new strategic actions where possible.
- Focus on cost-effective actions.
- Review treatment strategies.
- Review existing risk controls to determine if existing controls are not sufficient.
- Identify changes in thinking or new measures to overcome gaps (Marolla, 2017).

Ongoing Monitoring and Review

Information about the intensity and impact of the cyber-attack is continuously updated. Therefore, the risk assessment process includes monitoring and reviewing as important parts of the framework. This process should include the following steps:

1. Obtain new information as it becomes available.
2. Check that controls are effective.
3. Assess new information obtained from events.
4. Account for any changes in the process.
5. Identify new risks and take action (Marolla, 2017).

Societal Security: ISO 22301 – Continuity Management Approach

What would happen if an exponential surge in cyber-attacks to your organization, city, or even on your nation's vital infrastructure is happening right now? Would your entities survive the crisis? How would you ensure that you will endure the disaster? Business Continuity Management (BCM) is about preparing an organization/nation to deal with disruptive incidents that might otherwise prevent it from achieving its objectives. The international standard ISO 22301 specifies requirements to plan, establish, implement, operate, monitor, review, maintain, and continually improve a documented management system to protect against, reduce the likelihood of, prepare for, respond to, and recover from disruptive incidents when they arise (ISO 22301 Societal Security, 2017). This international standard can significantly reduce risk, particularly when a lack of awareness and a concrete strategy in continuity management planning after a disaster or impact is present. It also provides flexibility of implementation, identifying what is most relevant to minimize financial loss, infrastructure, communications, and the overall operation of the entity being attacked.

Almost every nation and/or establishment will be able to handle a crisis by the corresponding emergency services. But do we know what will happen next? Managing the overall continuity of operation becomes a priority in any type of severe weather event, as do implementing and operating controls and executing strategies for treating those risks. Any number of events can bring the city grinding to a halt and business continuity management planning will ensure leaders will respond judiciously to the circumstances. A continuity management approach contributes to a more resilient society (Societal Security Emergency Management, 2011; Marolla, 2017).

The following key components are considered when developing and implementing a continuity management system plan:

- Policy
- People with defined responsibilities
- Management processes relating to:
    - policy
    - planning
    - implementation and operation
    - performance assessment
    - management review
    - improvement.
- Documentation providing auditable evidence
- Any business continuity management processes relevant to the organization (Societal Security Emergency Management, 2011; Marolla, 2017).

Resuming "Normal Operations" after the Disaster

There are seven important points of action that determine how the nation or organization will preserve or restore critical functions. The nation's quick resumption of normal operations after a cyber-attack will affect the entire recovery.

1. *Establish an emergency planning team*: Federal, local government agents, and organizations' workers from all levels and departments must be included in the team, focusing on those with expertise vital to the prompt recovery of operations.
2. *Identify who is in charge*: It is important to identify who is in charge during a disaster risk event and ensure that all employees know

who that person or position is. Establish a procedure for the succession of management if that position or leader is not available at the time of the cyber-attack.

3. *Examine the nation-state of the situation or the organization operation and activities*: Identify internal and external operations that are important for the recovery and continuation of the different departments, municipalities, cities, telecommunication networks, and infrastructure.

4. *Identify an alternate location*: Important consideration is necessary to identify a different location to run the nation's critical infrastructure and/or different jurisdictions where leaders can conduct operations. Develop collaboration and viable assistance with "like" departments or the Public Building Commission/Department to share facilities if necessary.

5. *Communication plan*: Identify how the leaders and other staff will be advised of the emergency plan and what communication devices will be used in the cyber-attack event. List the communication tools in order of preference, emphasizing the most effective way for communication to the least effective according to the type and level of impact.

6. *Emergency contact list*: The emergency contact list identifies how to contact staff in the event of an incident or an indirect impact caused by the cyber-attack, such as bombarding your networks with malware around the clock, infecting your networks with different forms of malware, or finding and compromising your weakest networks in addition to consequences of disrupting your network systems (e.g., flooding, power outage, etc.). This will include the technical means where the staff may receive different types of communications in order to respond and lead actions toward restoring and aiding the operations and safeguarding the well-being of residents.

7. *Write a plan*: Document and update your continuity management system plans at least once a year (Business Continuity Plan, 2009).

It is difficult to predict or analyze the type of activities that will be interrupted due to cyber-attacks and how these impacts will affect the economic and social security of the organization. A continuity management plan draws attention to the impact of disruption and identifies those activities where the need to focus on its survival is urgent. Hacking has evolved from an individual experience to a large and collaborative scale with complex outcomes. Therefore, a continuity management system plan assists leaders in recognizing what needs to be done to protect its residents, infrastructure, and financial stability. ISO 22301 business continuity management system may also be able to take advantage of opportunities that might otherwise be overlooked or considered to be too high a risk.

Adapting and Using the Framework

The effective response to and recovery from global cyber-attacks involve taking prior actions before the event strikes. A proactive approach must be established along with planning for the likelihood of an event that has the capacity to interrupt the nation's operations and impinge on the well-being of the population and financial institutions. It is necessary to emphasize the personal commitment of the leaders to dismiss the thought that "it won't happen here." These attacks are often motivated by politics and activism and aimed at corporate or government networks and services and, as previously mentioned, preparedness means being proactive and planning. That is the essence of efficient business continuity planning (Marolla, 2017).

List of Terms

Business continuity management (BCM): A framework for identifying an organization's risk of exposure to internal and external threats. BCM includes disaster recovery, business recovery, crisis management, incident management, emergency management, and contingency planning.

Circular economy: An alternative to a traditional linear economy (make, use, dispose) in which we keep resources in use for as long as possible, extract the maximum value from them while in use, then recover and regenerate products and materials at the end of each service life.

Cloud service: Any service made available to users on demand via the Internet from a cloud-computing provider's servers as opposed to being provided from a company's own on-premises servers.

Cyber-security: Measures taken to protect a computer or computer system (as on the Internet) against unauthorized access or attack.

Environmental sustainability: The maintenance of the factors and practices that contribute to the quality of the environment on a long-term basis.

Risk management: The process of identifying, assessing, and controlling threats to an organization's capital and earnings. These threats, or risks, could stem from a wide variety of sources, including financial uncertainty, legal liabilities, strategic management errors, accidents, and natural disasters.

Technology neutrality: It refers to the same regulatory principles that should apply regardless of the technology used.

9

Mechanism Design, Risk Mechanism Theory and Its Relation to Information and Communication Technologies, Risk Management and Sustainable Development

Green Mechanism Design

This chapter describes in detail how mechanism design, risk management, a comprehensive strategic planning using SAC (strategic adaptive cognition) and risk mechanism theory can be used to address environmental risks and attain sustainable development goals (SDGs). Mechanisms allow you to revise a goal and learn something from the process for continual improvement, which is crucial to developing the right approach and acquiring the desired outcome. I named "Green Mechanism Design" as the strategic approach, applying "mechanisms" to solve problems and utilizing information and communication technologies (ICTs) in the practice of sustainability. Strategic planning needs to be developed for any type of risk. Hence, when addressing challenges – such as working out which technologies can be used to reduce greenhouse gas (GHG) emissions, assessing the potential GHG savings from introduction of green technologies, facing barriers to introduction of green technologies, or discovering how policy-makers and regulators can promote and enable introduction of green technologies and the benefits of enabling standards – these mechanisms help us understand institutions as the solution to a planner's problem of achieving some objective or maximizing some utility function subject to incentive constraints.

Mechanism design is the science of designing rules of a game to achieve a specific outcome, even though each participant may be self-interested. This is done by setting up a structure in which each player has an incentive to behave as the designer intends (Mechanism Design: Some Definitions and Results, 2016). I would say: "Everything is a mechanism and mechanisms are universal." "Universal" stands for a wide range of applications that support continued improvement and efficient operations. Nevertheless, mechanism enhances the existing applications and strategies and helps to cope with uncertainty. We develop our society through a series of mechanisms or strategic sets aimed at accomplishing a desirable outcome that has a social function. The social function needs to be implemented (the mechanism implements social function) which provides specific benefits for society as a whole. SDG challenges and outcomes are a characteristic of mechanism design. We need to take into consideration that without a strategy those goals will never be achieved.

If all of the agents' preferences were public knowledge, there would be no need for mechanism design – all that would need to be done is to solve the outcome optimization problem. Techniques from mechanism design are useful and necessary only in settings in which agents have private information about their preferences (Sen, 2007). Addressing global risks such as climate change, environmental degradation, resource scarcity, and economic disruptions that vary according to different factors related to local and international influences, present "unknown information" among the agents (nations, organizations, policy-makers, etc.). Mechanisms help to identify opportunities to solve problems and in this case the capability of nations to develop and implement strategies to accomplish SDGs. Overcoming its risks calls for the need for "mechanisms" or strategies to continue the processes of aligning the internal capabilities of the organization/nations with the external demands of its environment. In mechanism design you design the game; you formulate the set of guidelines that will provide a better understanding of the situation, identifying its obstacles for implementation; and you solve the problem. As a strategic approach to management, it involves the formulation and implementation of strategies to achieve the desired outcome. When you design a plan, you are designing a mechanism to attain your vision and therefore, achieve an optimal solution.

Nobel Laureate in Economics Dr. Eric Maskin referred to mechanism design as the reverse engineering part of the economics (Maskin, 2014). My focus is applying mechanism design not just as an economic tool to efficiently have an economic advantage or incentives toward desired objectives, but mainly a strategic setting framework that, along with sustainable and risk management frameworks, will result in a forward-thinking mindset approach for leaders to resolve any problems and in this case, problems related to environmental risk and sustainable development. The reason mechanisms (strategies) are universal is because we design mechanisms applicable to every aspect of our life in any situation, and how those mechanisms or series of systems and strategic moves (which are concerned with the incentives that must be applied to any set of agents that interact and exchange goods) influence decisions, outcomes, and society as a whole. Information and communication technology (ICT) is a tool that assists with the behavior we can build to create concrete change. Therefore, it is a mechanism that could provide a way to rethink and redesign the systems and processes leading to efficiency, influencing and shaping the behavior of participants. Mechanism design is the most effective way to create a systematic method to solve human problems that have definite consequences from the outcome. Therefore, paraphrasing Professor Maskin, "In mechanism design we start by identifying the outcome we would like to have and then we work backwards to figure out what institutions will generate those outcomes." A sustainable development agenda has specific goals leading to results (the 17 SDGs) that will change the shape of the world; therefore, the development and implementation of strategies and mechanisms that cope with climatic risks and provides a roadmap for climate actions will reduce emissions and build climate resilience, which is a crucial element in delivering concrete solutions and consequently accomplishing the aforementioned goals.

Strategy and Execution: Mechanism Design for Sustainable Development

Sustainable development requires a systemic response involving transformative changes, especially in policy and institutional systems from all sectors of society. SDGs and the Paris Agreement are global multi-stakeholder responses to this challenge and it calls for a universal framework that involves interdependent and contributing systems and trade-offs, just as an organization's outcomes involve trade-offs between multiple capitals. It is particularly relevant in the context of environmental and sustainability decision-making. Sustainable development is a normative concept which involves trade-offs among social, ecological, and economic goals, and is required to sustain the integrity of the overall system (Hediger, 2000).

Mechanism design, Professor Maskin explained, is the "reverse engineering part of economics." It can address potential trade-offs and inconsistencies among economic, social, and environmental policy objectives. You start with the goals that you want to achieve and then you work backwards to figure out what kind of procedure or mechanism will achieve those goals. Because SDGs are already established, mechanism design works perfectly with the development of "ways" to achieve such goals. Therefore, in regard to sustainable development, the goals are "predetermined" so we need to work on the different mechanisms to accomplish those goals. Mechanism design provides a framework for probing the inextricable connections between different agents' decision problems and our own. Therefore, the designed set of rules and processes (mechanisms) can support an already established system and/or develop a new mechanism to efficiently strategize coping with environmental risks. The main point would be to select the choice that maximizes the sum of agents' values with positive results benefiting the overall objectives. This is usefully formalized in terms of a social welfare function, which is based on an aggregate of individual preferences and, as a prerequisite of intergenerational equity and overall system integrity, on a set of sustainability constraints. In terms of addressing the well-being of the entire society, which is a component of sustainable development, social welfare is concerned with the quality of life that includes factors such as the quality of the environment. Hence, there is a case for strong environmental protection as consumption is competitive; that is, people care not only about their own consumption, but how their level of consumption compares to the consumption of their peers.

Environmental Pollution: A Threat to Sustainable Development

Environmental pollution is a threat to sustainable development. There are several environmental problems facing both developed and developing nations and are closely related to the welfare of human development. These include:

1. chemical pollution,
2. climate change,
3. resource and energy depletion, and
4. the loss of biodiversity and ecosystem integrity.

While often addressed separately, all four of these environmental problems are related to advancing industrialization, population growth, and the globalization of production and commerce (Ashford and Caldart, 2008). Pollution reduction strategies and regulations have not been very effective because they are designed using asymmetric and incomplete information. Furthermore, environmental burdens are often felt unequally within nations, between nations, and between generations, giving rise to international and intergenerational equity concerns that lead to concerns about environmental justice and equality.

For example, polluting firms can be better informed than the regulator with regard to the costs of reducing their emissions and the regulator may not have appropriate information about their own abatement costs. The information acquired by the regulators become significant in order to achieve pollution reductions and address the overall health risks for the population. In the following segment, Professor Eric Maskin's "Notes on Auctions for Pollution Reduction" provides a concrete example with solutions addressing firms that emit polluting GHGs:

Notes on Auctions for Pollution Reduction *Eric Maskin*

Suppose that there are n firms that emit polluting greenhouse gasses.

For each $i = 1, \ldots, n$, let $c_i = $ firm i's cost per unit of emissions reduction (*up to some maximum possible reduction.*)

Assume that $c_i \in [0, 1]$ and that c_i is private information for firm i.

If firm i reduces pollution by q_i and receives a monetary payment t_i, its overall payoff is:

$$t_i - c_i q_i$$

From an *ex ante* standpoint, each c_i can be thought of as an independent draw from a probability distribution. For the most part, we will assume that the distribution is discrete, but for some results we will consider a continuous distribution. The government has a budget B (which cannot be exceeded) for inducing voluntary reduction by firms.

PROBLEM:

What incentive-compatible and individually rational rule for allocating the money will induce the greatest expected reduction in pollution (where the expectation is taken with respect to the distribution)?

By individual rationality (IR), we mean that, if a firm bids truthfully, its payoff is nonnegative, i.e., $t_i - c_i q_i \geq 0$ for all possible realizations of t_i and q_i (this is actually *ex post* IR; a less demanding version of IR – interim IR – requires only that a firm's expected payoff be nonnegative, where the expectation is taken over other firms' equilibrium play).

In the analysis below, we shall, for the most part, require dominant strategy incentive-compatibility, i.e., that truthful bidding constitutes a dominant strategy. But we would obtain the same results if we demanded only Bayesian incentive-compatibility. On the other hand, the fact that we require *ex post* IR is important.

In fact, as we will show, the confluence of three assumptions – (A) *ex post* IR, (B) a maximum possible reduction that is finite, and (C) a fixed budget for the government – is what makes our solution novel. If any one of these assumptions were replaced by a more "standard" hypothesis, the optimal allocation rule would reduce to a standard mechanism.

To see this, let us begin by seeing what happens if, instead of having a fixed budget (assumption C), the government has an unlimited budget that it can use to promote social welfare.

Formally, the government maximizes the gross social benefit of pollution reduction, $V(\Sigma_i^n q_i)$, less the cost of reduction, $\Sigma_{i-1}^n c_i q_i$. This problem reduces to the one we begin with, that of maximizing pollution reduction, if, when we impose budget constraint $\Sigma t_i \leq B$, the constraint is binding. To reflect the idea that there is a limit to how big reductions can be (assumption B), let us suppose that, $q_i \leq 1$ for all i (we assume here that all firms face the same limit, but this is not necessary). Then the government can attain the optimum by running a Groves mechanism. Specifically, we have:

Proposition 1:

Suppose that the government wishes to maximize

$$V\left(\sum_{i=1}^{n} q_i\right) - \sum_{i=1}^{n} c_i q_i$$

where $V(0)=0$, using a mechanism that satisfies *ex post* IR, dominant-strategy, incentive-compatibility, and the constraint that

$$q_i \leq 1 \text{ for all } i$$

Then the government can attain the optimum via the following procedure:

(i)   Each firm i makes us a bid $\hat{c}_i$

(ii) Firm i is required to reduce by an amount q_i^*, where

$$\left(q_1^*,\ldots,q_n^*\right)= \arg \begin{array}{c} \max \\ q, \leq 1 \\ j=1,\ldots,n \end{array} \left(V\left(\sum_{j=1}^{n} q_j\right) - \sum_{j=1}^{n} \hat{c}_j q_j\right)$$

(iii) Firm i is paid an amount

$$t_i = V\left(\sum_{j=i}^{n} q_j^*\right) - \sum_{j \neq i}^{n} \hat{c}_j q_j$$

Proof: Because of (iii), firm i's objective function

$$(*)V\left(\sum_{j=1}^{n} q_j^*\right) - \sum_{j \neq i}^{n} \hat{c}_j q_j^* - c_i q_i^*$$

coincides with the social maximand. Hence, truthful bidding constitutes a dominant strategy. By construction, the mechanism attains the optimum. Because $V(0)=0$, (*) is nonnegative if firm i is truthful, and so the mechanism satisfies *ex post* IR (assumption A). Because $q_j \leq 1$ in (ii), it adheres to the upper bound on pollution reductions (assumption B).

Q.E.D. ("that which was to be demonstrated")

   There is, however, no reason why

$$\sum t_i$$

should not exceed B, where t_i is defined by (iii) (i.e., assumption C may be violated). Thus, the Groves mechanism is not a solution to the problem.

Let $\pi(c_i) = probability\ of\ c_i$. The solution to our problem turns out to depend critically on whether $\pi(c_i)$

$$\pi(c_i) \text{ is increasing in } c_i \tag{9.1}$$

This means that high costs are more likely than low costs. In the case of continues distribution, this corresponds to the assumption that

$$F'' \geq 0, \tag{9.2}$$

where F is the cumulative distribution function. Let us stick with a continuous distribution for the time being and assume, in addition to (9.2) that

$$\frac{d}{dx}\left(\frac{F(x)}{xF'(x)} \right) \geq 0 \tag{9.3}$$

Given conditions (9.2) and (9.3), if we replace assumption B with the supposition that it is not upper bound to pollution reductions, then, once again, the optimal rule will take a standard form- specifically, it reduces to a second-price auction.

Proposition 2:

Assume that (9.2) and (9.3) hold. Suppose that, for all i, the maximum reduction q_i that firm i can make is unbounded. Then, the Bayesian incentive-compatible and (interim) individual rational budget-allocation rule that, in equilibrium, maximizes the expected pollution reduction takes the following form:

(i)   Each firm i makes a bid $\hat{c}_i$
(ii) The lowest bidder (i.e., the bidder i^* such that $\hat{c}_{i^*} = \min, _i\hat{c}_i)$ is declared the winner
(iii) The winner is paid a price per unit reduction p equal to the second-lowest bid, i.e.,

$$p = \min_{i \neq i^*} \hat{c}_i$$

(iv) The winner reduces by an amount q that just exhaust the budget $B(\geq 1)$ at price p, i.e.,

$$q = \frac{B}{p}.$$

Proof:

$$\text{Let}_i = \frac{B}{c_i} \text{ and } T_i = \frac{t_i}{B}$$

Then we can rewrite firm i's maximand as

$$E[\theta_i T_i - q_i] \tag{9.4}$$

Suppose we think of $\sum_{i=1}^{n} q_i$ as the auctioneer's "revenue" T_i as the quantity of "goods" sold to firm i. Then $\sum T_i \leq 1$ corresponds to the constraint that the auctioneer can sell no more than one unit of the goods, and the auctioneer's problem can be reinterpreted as the standard one of maximizing expected revenue.

From Myerson (1981) and Riley and Samuelson (1981), we know that the form of the optimal auction depends on the behavior of

$$J(\theta) = \theta - \frac{1 - F_0(\theta)}{F_0'(\theta)}$$

where

$$F_0(\theta) = 1 - F\left(\frac{B}{\theta}\right), \quad \theta \in [B, \infty]. \tag{9.5}$$

Taken $x = 1/\theta$. Then from (9.5),

$$J\left(\frac{1}{x}\right) = \frac{1}{x} - \left(\frac{FBx}{BF'(Bx)x^2}\right) \tag{9.6}$$

Formula (9.6) implies that

$$-J^{\left(\frac{1}{x}\right)\frac{1}{x^2}} = \frac{-2}{x^2} + \frac{FF''}{(F')^2 x^2} + \frac{2F}{BF'x^3} \tag{9.7}$$

From (9.3) we have

$$\frac{2}{x^2} - \frac{2FF''}{(F')^2 x^2} - \frac{2F}{F'x^3} \geq 0 \tag{9.8}$$

Adding the left-hand side of (9.8) to the right-hand side of (9.7), we obtain

$$\frac{-FF''}{(F')^2 x^2} + \frac{2F}{F'x^3}\left(\frac{1}{B} - 1\right),$$

Which because $B \geq 1$, is negative from (9.2). Hence, the right-hand side of (9.6) is also negative, and we have

$$J'(x) > 0 \tag{9.9}$$

But the Myerson and Riley-Samuelson analysis than implies that the optimal T_i satisfies

$$T_i = \begin{cases} 1, & \text{if } i = i^* \\ 0, & \text{otherwise,} \end{cases}$$

where:

$$\theta_{i^*} = \max_j \theta_j,$$

$$p = \max_{i \neq i^*} \theta_i$$

implying that

$$t_i = \begin{cases} B, & \text{if } i = i^* \\ 0, & \text{otherwise} \end{cases}$$

and

$$q_i = t_i/p$$

Q.E.D.

Observe that the second-price auction of Proposition 2 satisfies *ex post* IR (assumption A) and also adheres to the fixed budget constraint (assumption C).

So far, we have seen how the optimal mechanism reduces to something familiar when, in turn, we eliminate assumption C or B or allow firms. To complete the preliminaries, let us examines what happens when we replace *ex post* IR with interim IR (i.e., relax assumption A). Chung and Ely (2002) show that, in that case, the optimal mechanism takes the form of a Baron-Myerson (1982) mechanism.

Proposition 3 (Chung and Ely (2002)):

Assume that (9.2) and (9.3) hold, and that for all i, the maximum reduction that a firm i can make is 1. Then, the optimal dominant-strategy incentive-compatible and interim IR mechanism has the property that there exists $c^* \leq 1$ such that, for all i, if $c_i \leq c^*$, firm i reduces by 1 and if $c_i > c^*$, firm i makes no reduction.

Proof: See Chung and Ely (2002).

When $n = 2$, $B = 1$, and the distribution of c_i is uniform, the mechanism of Proposition 3 becomes:

(i) Each firm i, $i = 1,2$, bids $\hat{c}_i$
(ii) Firm i reduces by

$$q_i = \begin{cases} 1, & \text{if } \hat{c}_i \leq \sqrt{\dfrac{1}{2}} \\ 0, & \text{if } \hat{c}_i > \sqrt{\dfrac{1}{2}} \end{cases}$$

(iii) Firm i is paid

$$t_i = \begin{cases} \dfrac{1}{2}, & \text{if } \hat{c}_i, \hat{c}_j \leq \sqrt{\dfrac{1}{2}} \text{ or } \hat{c}_i, \hat{c}_j > \sqrt{\dfrac{1}{2}} \\[2ex] \dfrac{1}{2} + \sqrt{\dfrac{1}{2}}, & \text{if } \hat{c}_i \leq \sqrt{\dfrac{1}{2}} \text{ and } \hat{c}_j > \sqrt{\dfrac{1}{2}} \\[2ex] \dfrac{1}{2} + \sqrt{\dfrac{1}{2}}, & \text{if } \hat{c}_i \leq \sqrt{\dfrac{1}{2}} \text{ and } \hat{c}_j > \sqrt{\dfrac{1}{2}} \end{cases}$$

Notice that it is a dominant strategy to bid truthfully in this mechanism. Furthermore, the budget is always exactly exhausted (even if there are no reductions). However, the mechanism clearly violates *ex post* IR. If, for example, $\hat{c}_i > \sqrt{\frac{1}{2}}$ but $\hat{c}_j \cdot \sqrt{\frac{1}{2}}$, then the firm i will be forced to pay $\sqrt{\frac{1}{2} - \frac{1}{2}}$, which leaves it worse off than had it not participated.

Henceforth, let us use a discrete distribution and suppose that the support of π is $\left\{0, \frac{1}{m}, \cdots, \frac{m}{m}\right\}$, for some integer m. For now, we will suppose that each firm can reduce by a maximum of 1 unit (we will consider the case of heterogeneous capacity below). If $B > n$, we don't need an auction: The auctioneer can simply offer to pay any firm at a rate of 1 per unit reduction. (Since all costs in the support of π are less than 1, each firm will be willing to reduce up to its capacity. And, since $n < B$, the auctioneer will remain within the budget even with maximal reduction.)

We shall assume, therefore, that $B < n - 1$. We have:

Proposition 4:

Suppose that $n = 2$ and $B < 1$. Assume that (9.1) holds and that each firm can reduce pollution by a maximum of 1 unit. Suppose also that there exists $K > 1$ such that $\pi\left(\frac{i+1}{m}\right) \le K\pi\left(\frac{i}{m}\right)$ for all i. Then, for K sufficiently near 1 the dominant-strategy incentive-compatible and *ex post* individually-rational budget-allocation rule that, in equilibrium, maximizes expected pollution reduction, taking the following form:

(i) Each firm imakes a bid $\hat{c}_i$

(ii) If firm is the lower bidder $\hat{c}_i < \hat{c}_j$, then it reduces by q_i, where

$$q_i = \begin{cases} 1, & \text{if } \hat{c}_j \le B \\ \dfrac{B}{\hat{c}_j}, & \text{if } \hat{c}_j > B \end{cases}$$

and is paid an amount t_i, where

$$t_i = \begin{cases} \dfrac{B}{2} + \dfrac{\hat{c}_j^2}{2B}, & \text{if } \hat{c}_j \le B \\ B, & \text{if } \hat{c}_j > B \end{cases}$$

(iii) If firm is the higher bidder $\hat{c}_i > \hat{c}_j$, then it reduces by q_i, where

$$q_i = \begin{cases} 1 - \dfrac{\hat{c}_i}{2B}, & \text{if } \hat{c}_i \le B \\ 0, & \text{if } \hat{c}_i > B \end{cases}$$

and is paid an amount t_i, where

$$t_i = \begin{cases} \dfrac{B}{c} - \dfrac{\hat{c}_i^2}{2B}, & \text{if } \hat{c}_i \leq B \\ 0, & \text{if } \hat{c}_i > B \end{cases}$$

(iv) If $\hat{c}_i = \hat{c}_j$, then

$$q_i = \begin{cases} 1 - \dfrac{\hat{c}_i}{2B}, & \text{if } \hat{c}_i = \hat{c}_j \leq B \\ \dfrac{B}{2\hat{c}_i}, & \text{if } \hat{c}_i = \widehat{c_j > B} \end{cases}$$

and $t_i = (B/2)$

Proof: We can write the maximization problem as

$$\text{Max} \sum_i \sum_j \big(q(i,j) + q(j,i) \big) \pi(i) \pi(j) \tag{9.10}$$

subject to

$$t(i,j) - \frac{i}{m} q(i,j) \geq t(k,j) - \frac{i}{m} q(k,j), \quad \text{for all } i, k, j, \tag{9.11}$$

$$t(i,j) - \frac{i}{m} q(i,j) \geq 0, \text{ for all } i \text{ and } j, \tag{9.12}$$

$$q(i,j) \leq 1, \text{ for all } i \text{ and } j, \tag{9.13}$$

$$q(i,j) \geq 0, \text{ for all } i \text{ and } j, \tag{9.14}$$

and

$$t(i,j) + t(j,i) \leq B \tag{9.15}$$

where $q(i,j)$ and $t(i,j)$ are, respectively, the reduction made, and payment received by firm i with cost i/m if the other firm's cost is j/m (because of symmetry, we can take the q and t functions to be the same for both firms), and $\pi(i)$ is shorthand for $\pi(i/m)$.

We first argue that the incentive constraints (9.11) can be replaced by

$$t(i,j) - \frac{i}{m} q(i,j) \geq t(i=1,j) - \frac{i}{m} q(i+1,j), \quad \text{for all } i, j, \tag{9.16}$$

and

$$q(i,j) \geq q(i+1,j), \text{ for all } i, j. \tag{9.17}$$

That is, rather than imposing all incentive constraints, we need require only the "upward and adjacent" constraints (9.16), as long as the monotonicity requirement (9.17) is invoked. Furthermore, we can replace the individual–rationality constraints (9.12) with the weaker requirement

$$t(1,j) - q(1,j) \geq 0 \text{ for all } i \text{ and } j. \tag{9.18}$$

To see this, consider the program (9.10), and (9.13)–(9.18). We claim that there is a solution in which (9.16) and (9.18) are binding, for all i and j. If not, consider a solution in which, for some i and j,

$$t(i,j) - \frac{i}{m}q(i,j) > t(i+1,j) - \frac{i}{m}q(i+1,j). \tag{9.19}$$

But then we can reduce t (i, j) until equality holds without violating any constraint.

Furthermore, we can do this for all i and j for which (9.19) holds. A similar argument applies if the inequality (9.18) holds strictly. We may assume, therefore, that the solution to program (9.10), (9.13)–(9.18) satisfies that

$$t(i,j) - \frac{i}{m}q(i,j) = t(i+1,j) - \frac{i}{m}q(i+1,j) \text{ for all } i \text{ and } j \tag{9.20}$$

and

$$t(1,j) - q(1,j) = 0 \tag{9.21}$$

But from (9.17), (9.20) implies that

$$t(i,j) - \frac{k}{m}q(i,j) \geq t(i+1,j) - \frac{k}{m}q(i+1,j) \text{ for all } i, j, \text{ and } k \leq i. \tag{9.22}$$

Applying (9.22) repeatedly, we have

$$t(k,j) - \frac{k}{m}q(k,j) \geq t(l,j) - \frac{k}{m}q(l,j) \text{ for all } k \text{ and } l \geq k. \tag{9.23}$$

From (9.17) and (9.20) we have

$$t(i,j) - \frac{k}{m}q(i,j) \leq t(i+1,j) - \frac{k}{m}q(i+1,j) \text{ for all } i, j \text{ and } k > i. \tag{9.24}$$

Applying (9.24) repeatedly we obtain

$$t(l,j) - \frac{k}{m}q(l,j) \leq t(k,j) - \frac{k}{m}q(k,j) \text{ for all } k \text{ and } l \leq j. \tag{9.25}$$

But (9.23) and (9.25) together constitute the full set of inequalities in (9.11). Finally, note that (9.18) and (9.23) imply that

$$t(k,J) - \frac{k}{m}q(k,j) \geq 0 \text{ for all } k \text{ and } j.$$

Now, (9.20) and (9.21) can be expressed as

$$t(i,j) = \frac{i}{m}q(i,j) + \frac{1}{m}\sum_{k=i+1}^{m}q(k,j). \tag{9.26}$$

From (9.26) we can rewrite the constraint (9.15) as

$$\frac{1}{m}\left(iq(i,j) + \sum_{k=i+1}^{m} q(k,j) + jq(i,j) + \sum_{k=j+1}^{m} q(k,i) \right) \leq B \qquad (9.27)$$

Consider the program (9.10), (9.13), (9.14), (9.17), and (9.27). From the above analysis, we know that a solution also solves the original program. In fact, we will consider the reduced program in which (9.17) is omitted and show that a solution to that program turns out to satisfy the omitted constraint.

Consider the first-order conditions of the reduced program with respect to $q(i, j)$:

$$2\pi(i)\pi(j) - \sum_{k=1}^{i-1} \alpha_{kj} \frac{1}{m} - \alpha_{ij} \frac{i}{m} - \beta_{ij} + \gamma_{ij} = 0, \text{ if } i < j \qquad (9.28)$$

$$2\pi(i)\pi(j) - \sum_{k=1}^{nj-1} \alpha_{kj} \frac{1}{m} - \frac{2}{m}\alpha_{jj} - \sum_{k=j+1}^{i-1} \frac{1}{m}\alpha_{jk} - \frac{i}{m}\alpha_{ji} - \beta_{ij} + \gamma_{ij} = 0, \text{ if } i > j, \quad (9.29)$$

and

$$2(\pi(i))^2 - \sum_{k=1}^{i-1} \alpha_{ki} \frac{1}{m} - \frac{2i}{m}\alpha_{ii} - \beta_{ii} + \gamma_{ii} = 0 \qquad (9.30)$$

where:

α_{ij} is the Lagrange multiplier for constraint (9.27)
β_{ij} is the Lagrange multiplier for constraint (9.13)
γ_{ij} is the Lagrange multiplier for constraint (9.14)

Because the program is linear, and the $q(i, j)$s and $t(i, j)$s of the Proposition satisfy (9.13), (9.14), (9.17), (9.26), and (9.27), it will suffice to show that there exists a nonnegative solution $(\alpha_{ij}, \beta_{ij}, \gamma_{ij})$ to (9.28)–(9.30) such that, for all i and j, the following complementary slackness conditions hold for these $q(i, j)$s:

$$\beta_{ij}(1 - q(i,j)) = 0, \qquad (9.31)$$

$$\alpha_{ij}\left(B - \frac{i}{m}q(i,j) - \frac{1}{m}\sum_{k=i+1}^{m} q(k,j) - \frac{j}{m}q(j,i) - \frac{1}{m}\sum_{k=j+1}^{m} q(k,i) \right) = 0, \qquad (9.32)$$

and

$$\gamma_{ij}q_{ij} = 0. \qquad (9.33)$$

To show that such a solution $(\alpha_{ij}, \beta_{ij}, \gamma_{ij})$ exists, we will begin with the case in which $j < i \leq Bm$.

Take

$$\beta_{im} = \gamma_{ij} = 0 \tag{9.34}$$

Then, from (9.29) and (9.34), we have

$$\alpha_{ji} = \frac{2m}{i}\pi(i)\pi(j) - \sum_{k=1}^{j-1}\alpha_{kj}\frac{j-1}{i(i-1)} - \sum_{k=j}^{i-1}\pi(j)\pi(k)\frac{2m}{i(i-1)}. \tag{9.35}$$

We claim $\alpha_{ji} \geq 0$. to see this note first that

$$\alpha_{12} = m\left(\pi(1)\pi(2) - \pi(1)^2\right) \geq 0.$$

Suppose, that the claim is false and i is the smallest integer such that there exists $j < i$ with $\alpha_{ij} < 0$. Then, from the choice of i, $\alpha_{kj} \geq 0$ for all $k < j$.
Then, from (9.35), we have, for $k < j \leq Bm$,

$$\alpha_{kj} \leq \frac{2m\pi(k)\pi(j)}{j} \tag{9.36}$$

From (9.35) and (9.36) and because π is nondecreasing,

$$\alpha_{ji} \geq \frac{2m}{i}\pi(i)\pi(j) - (j-1)^2\pi(k)\pi(j)\frac{2m}{ji(i-j)} - (i-j-1)\pi(j)\pi(i)\frac{2m}{i(i-1)}$$

$$\geq \frac{2m}{i}\pi(i)\pi(j) - (i-2)\pi(i)\pi(j)\frac{2m}{i(i-1)} \geq \pi(i)\pi(j)\frac{4m}{i(i-1)} > 0$$

contradicting our choice of i. We conclude that $\alpha_{ji} \geq 0$ for all $j < i...Bm$, as claimed. Next, consider the case $i = j...Bm$. Take

$$\beta_{ii} = \gamma_{ii} = 0 \text{ for all } i \leq Bm \tag{9.37}$$

From (9.35) and because π is nondecreasing, we have, for $k < i...Bm$;

$$\alpha_{ki} \leq \frac{2m}{i}\left(\pi(i)\right)^2 \tag{9.38}$$

Hence, from (9.38), we have

$$2\left(\pi(i)\right)^2 - \sum_{k=1}^{i-1}\alpha_{ki}\frac{1}{m} \geq 0 \text{ for all } i \leq Bm \tag{9.39}$$

Thus, from (9.37) and (9.39), if we choose α_{ii} to satisfy (9.30), α_{ii} will be nonnegative, as required.
Next, consider $i < j...Bm$. From (9.29) and the above analysis, we have

$$2\pi(j)\pi(i) - \sum_{k=1}^{i-1}\alpha_{ki}\frac{1}{m} - \frac{2}{m}\alpha_{ii} - \sum_{k=i+1}^{j-1}\frac{1}{m}\alpha_{ik} - \frac{j}{m}\alpha_{ij} = 0 \tag{9.40}$$

In order to conclude that, for all $i < j < Bm$, we can take $\beta_{ij} > 0$ and $\gamma_{ij} = 0$, we must show, in view of (9.28), that

$$2\pi(i)\pi(j) - \sum_{k=1}^{i-1}\alpha_{kj}\frac{1}{m} - \frac{i}{m}\alpha_{ij} > 0 \qquad (9.41)$$

But, because $i < j$, (9.41) follows from (9.40) provided that

$$\alpha_{ki} \geq \alpha_{kj}, \text{ for all } k < i < j \leq Bm \qquad (9.42)$$

From (9.35),

$$\alpha_{ki} = \frac{2m}{i}\pi(k)\pi(i) - \sum_{l=1}^{k-1}\alpha_{lk}\frac{k-1}{i(i-1)} - \sum_{l=k}^{i-1}\pi(k)\pi(l)\frac{2m}{i(i-1)} \qquad (9.43)$$

if $K = 1$ then $\pi(j) = \pi$ for all j, and so, for $k < i$, (9.43) can be rewritten as

$$\alpha_{ki} = \frac{k-1}{i(i-1)}\left(2m\pi^2 - \sum_{l=1}^{k-1}\alpha_{lk}\right)$$

Similar, for $k < j$, we have

$$\alpha_{kj} = \frac{k-1}{j(j-1)}\left(2m\pi^2 - \sum_{l=1}^{k-1}\alpha_{lk}\right)$$

and so because $>$, (9.42) holds, as claimed, for $K > 1$ sufficiently near 1.
Finally, consider i, j such that $j > Bm$. For $i < j$, choose

$$\beta_{ij} = \gamma_{ij} = 0 \qquad (9.44)$$

From (9.28) and (9.44), we have

$$\alpha_{1j} = 2m\pi(1)\pi(j), \text{ for } j > Bm \qquad (9.45)$$

and

$$\alpha_{ij} = \frac{2m}{i}\pi(i)\pi(j) - \frac{2m}{i(i-1)}\sum_{k=1}^{i-1}\pi(k)\pi(j), \text{ for all } 2 \leq i\langle j, j\rangle Bm \qquad (9.46)$$

But because π is nondecreasing,

$$\frac{2m}{i}\pi(i)\pi(j) - \frac{2m}{i(i-1)}\sum_{k=1}^{i-1}\pi(k)\pi(j) \geq \frac{2m}{i}\pi(i)\pi(j) - \frac{2m}{i}|pi(i)\pi(j) = 0, \text{ for } i > 2$$

and so, from (9.45) and (9.46), $\alpha_{kj} \geq 0$. Next, consider $i = j > Bm$. Again, choose β_{ij} and γ_{ij} to satisfy (9.44). From (9.30), (9.45), and (9.46), we have

$$2(\pi(i))^2 - \sum_{k=1}^{i-1}\alpha_{ki}\frac{1}{m}\begin{cases} 2(\pi(1))^2, & i = 1 \\ 2(\pi(i))^2 - \frac{2}{i-1}\sum_{k=1}^{i-1}\pi(k)\pi(i), i \geq 2 \end{cases} \qquad (9.47)$$

Because π is nondecreasing, the right-hand side of (9.47) for $i \geq 2$ is nonnegative.

Hence, if we take

$$\alpha_{ii} = \frac{m}{2i} \left(2\left(\pi(i)\right)^2 - \sum_{k=1}^{i-1} \alpha_{ki} \frac{1}{m} \right) \geq 0 \tag{9.48}$$

we will satisfy (9.30). Last, consider $i > j > Bm$. If $K = 1$, then $\pi(k) = \pi$ for all k, and so (9.45) and (9.46) imply that

$$\alpha_{kj} = \begin{cases} 2m\pi^2, & \text{for } k = 1, \ j > Bm \\ 0, & j > k \geq 2, \ j > Bm \end{cases} \tag{9.49}$$

Hence, from (9.48) and (9.49),

$$2\pi^2 - \sum_{k=1}^{j-1} \alpha_{kj} \frac{1}{m} - \frac{2}{m} \alpha_{jj} - \sum_{k=j+1}^{i-1} \alpha_{jk} \frac{1}{m} - \frac{i}{m} \alpha_{ji} \geq 0 \tag{9.50}$$

Thus, if we take $\beta_{ij} = 0$ and γ_{ij} to satisfy (9.29), (9.50) implies that $\gamma_{ij} \geq 0$
Because $q(i, j) = 0$ we conclude that (9.32) holds.

Q.E.D.

We can readily extend Proposition 4 to three or more bidders. Here is how the auction looks in the case $n = 3$.

Proposition 5:

Assume that the hypotheses of Proposition 4 hold. If $n = 3$, then, for K sufficiently near 1, the optimal dominant-strategy incentive-compatible and *ex post* IR mechanism takes the following form:

(i) Each firm i makes a bid $\hat{c}_i$; if $\hat{c}_i < \hat{c}_j < \hat{c}_k$, then:
(ii) Each firm (the lower bidder) reduces q_i, where

$$q_i = \begin{cases} 1, & \text{if } \hat{c}_j \leq B \\ \dfrac{B}{\hat{c}_j}, & \text{if } \hat{c}_j > B \end{cases}$$

and is paid an amount t_i, where

$$t_i = \begin{cases} \dfrac{B}{2} + \dfrac{4}{3} \dfrac{\hat{c}_k^3}{B^2}, & \text{if } \hat{c}_k < \dfrac{B}{2} \\[2ex] \dfrac{B}{2} - \dfrac{B\hat{c}_j^2}{2\hat{c}_k^2} + \dfrac{\hat{c}_j^2}{\hat{c}_k}, & \text{if } \dfrac{B}{2} < \hat{c}_k < B \\[2ex] \dfrac{B}{2} + \dfrac{\hat{c}_i}{2B}, & \text{if } \hat{c}_j < B < \hat{c}_j \\[2ex] B, & \text{if } \hat{c}_j > B \end{cases}$$

(iii) Firm j (the middle bidder) reduces by q_i, where

$$q_j = \begin{cases} 1, & \text{if } \hat{c}_k < \dfrac{B}{2} \\[2mm] 1 + \dfrac{B\hat{c}_j}{\hat{c}_k^2} - \dfrac{2\hat{c}_j}{\hat{c}_k}, & \text{if } \dfrac{B}{2} < \hat{c}_k < B \\[2mm] 1 - \dfrac{\hat{c}_j}{B}, & \text{if } \hat{c}_j < B < \hat{c}_k \\[2mm] 0, & \text{if } \hat{c}_j > B \end{cases}$$

and is paid

$$t_j = \begin{cases} \dfrac{B}{3} + \dfrac{4\hat{c}_k^3}{B^2}, & \text{if } \hat{c}_k < \dfrac{B}{2} \\[2mm] \dfrac{B}{2}\left(1 - \dfrac{\hat{c}_j^2}{B^2}\right), & \text{if } \hat{c}_j < B < \hat{c}_k \\[2mm] 0, & \text{if } \hat{c}_j > B \end{cases}$$

(iv) Firm k (the highest bidder) reduces by

$$q_k = \begin{cases} 1 - \dfrac{4\hat{c}_k^2}{B^2}, & \text{if } \hat{c}_k < \dfrac{B}{2} \\[2mm] 0, & \text{if } \dfrac{B}{2} < \hat{c}_k \end{cases}$$

and is paid

$$t_k = \begin{cases} \dfrac{B^2 - \hat{c}_k^3}{3B}, & \text{if } \hat{c}_k < \dfrac{B}{2} \\[2mm] 0, & \text{if } \dfrac{B}{2} < \hat{c}_k \end{cases}$$

Proof: Completely analogous to that of Proposition 4. Next, consider the case in which

$$\pi\left(\frac{i}{m}\right) \geq K\pi\left(\frac{i+1}{m}\right) \text{ for some } K \geq 1. \tag{9.51}$$

That is, costs are more likely to be low than high. There is some evidence that this was the relevant case in the British pollution auction.

Proposition 6:

Assume that (9.51) holds and that each firm can reduce by a maximum of 1 unit. For K big enough, the optimal dominant-strategy incentive-compatible and *ex post* IR mechanism takes the following form:

(i) Each firm I makes a bid $\hat{c}_i$
(ii) If there are no bids below B, then

$$q_1 = \cdots = q_n = 0$$

(iii) If, for some $l_e\{1,\ldots,n\}$, there are exactly l bids less than or equal to B/l, then firm reduces by

$$q_i = \begin{cases} 1, & \text{if } \hat{c}_i \le \dfrac{B}{l} \\[2mm] 0, & \text{if } \hat{c}_i > \dfrac{B}{l} \end{cases}$$

and is paid

$$t_i = \begin{cases} \dfrac{B}{l}, & \text{if } \hat{c}_i \le \dfrac{B}{l} \\[2mm] 0, & \text{if } \hat{c}_i > \dfrac{B}{l} \end{cases}$$

(iv) If, for some l, there are more than l bids less than or equal to B/l but fewer than $l+1$ less than equal to $B/(1 + 1)$, then firm i reduces by

$$q_i = \begin{cases} 1 & \text{if } \hat{c}_i < \hat{c}^{l+1} \\[2mm] \dfrac{(B/\hat{c}_i)-h}{k-h} & \text{if } \hat{c}_i = \hat{c}^{l+1} \text{ and there are } n-k \text{ bids strictly higher than } \hat{c} \\[1mm] & \text{and } h \text{ bids strictly lower than } \hat{c}_i, \text{ where } k > l > h \\[2mm] 0 & \text{otherwise} \end{cases}$$

and is paid

$$t_i = \hat{c}^{l+1} q_i$$

where $\hat{c}^{l+1}$ is the $(l+1)$st – lowest bid.

Proof: Generalizing the argument in the proof of Preposition 4, we can show that if

if $\left(q_i(\cdot),\ldots,q_n(\cdot)\right)$ solves the program

$$\max_{(q_1(\cdot),\ldots,q_n(\cdot))} \sum_{k_1=1}^{m} \cdots \sum_{k_n=1}^{m} \sum_{i=1}^{n} q_i(k_1,\ldots,K_n)\pi(k_1)\ldots\pi(k_n) \qquad (9.52)$$

subject to

$$0 \le q_i(k_1,\ldots,k_n) \le 1 \text{ for all } i \text{ and } (k_i,\ldots,k_n) \qquad (9.53)$$

and

$$\frac{1}{m}\left(\sum_{i=1}^{n}\left(k_i q_i(k_i,k_{-i}) + \sum_{s=k_i+1}^{m} q_i(s,k_{-i})\right)\right) \le B \text{ for all } (k_i,k_{-i}), \qquad (9.54)$$

where $q_i\,(k_i q_i(k_i,k_{-i}))$ is firm i's reduction if its cost $\dfrac{k_i}{m}$ and the vector of the other firms' costs is $\dfrac{k_{-i}}{m}$, then, provided that

$$q_i(k_i, k_{-i}) \geq q_i(k_i + 1, k_{-i}) \text{ for all } i, k_{-i}, \text{ and } k_{i \in \{1, \ldots, m-1\}}, \tag{9.55}$$

$(q_1(\cdot), t_1(\cdot)), \ldots, (q_n(\cdot), t_n(\cdot))$ constituted an optimal mechanism, where for all i and (k_i, k_{-i}),

$$t_i(k_i, k_{-i}) = \frac{1}{m}\left(k_i q_i(k_i, k_{-i}) + \sum_{s=k_i+1}^{m} q_i(s, k_{-i}) \right). \tag{9.56}$$

and $t_i(k_i, k_{-i})$ is the payment to firm i if its reduction is $q_i(k_i, k_{-i})$. It is a simple computation to verify that (9.53)–(9.55) are satisfied if q_i satisfies (ii)–(iv) of the Proposition. Thus, it suffices to show that (ii)–(iv) solve the program (9.52)–(9.54).

Suppose, in contradiction to this claim, that the solution violates one of (ii), (iii), and (iv). Consider the case in which (iii) fails to hold. Then, without loss of generality, we may assume that there exists

$$k_1 \leq k_2 \leq \cdots \leq k_l \leq \frac{Bm}{l} < k_{l+1} \leq \cdots \leq k_n$$

such that either

$$q_i(k_1, \ldots, k_n) < 1 \text{ for some } i \leq l \tag{9.57}$$

or

$$q_i(k_1, \ldots, k_n) > 0 \text{ for some } i > l. \tag{9.58}$$

Now if (9.58) holds, then for some

$$j > \frac{Bm}{l} \text{ and } k_1' \leq k_2' \leq \cdots \leq k_l' \leq \frac{Bm}{l} < k_{l+1}' \leq \cdots k_n',$$

$$q_j(k_1', \ldots, k_n') < 1,$$

or else one of the constraints (9.54) will be violated. Thus, we can assume that (9.57) holds.

Consider increasing (k_i, k_{-i}) by $\frac{\epsilon}{k_i}$. If this violates one of the constraints (9.54), then there must exist

$$j > \frac{Bm}{l} \text{ and } k_i' \leq k_i \text{ such that } q_j(k_i', k_{-i})$$

appears in that constraint and

$$q_j(k_i', k_{-i}) > 0$$

Hence, if we decrease $q_j(k_i', k_{-i})$ by ϵ for each such constraint, (9.54) will continue to hold. But because $k_i \leq \frac{Bm}{l} < k_j$ we have

$$\pi\left(\frac{i}{m}\right) > K\pi\left(\frac{j}{m}\right)$$

Thus, for K big enough, these changes in the q's increase the maximand, a contradiction. We conclude that (iii) must hold after all. A similar argument applies for (ii) and (iv).

Q.E.D.

Up to now we have been assuming that all firms have the same reduction capacity, i.e., that they can reduce up to 1 unit. We now relax this assumption and suppose that

$$\text{Firm } i \text{ can reduce up to } r_i, \text{ where } r_{i\varepsilon}[0,1] \qquad (9.59)$$

where the r_i's are independent draws from a discrete probability distribution $p(r)$

A firm i is now described by the pair (c_i, r_i).

Proposition 7:

If $\pi\left(\dfrac{i}{m}\right) / \pi\left(\dfrac{i+1}{m}\right) > k$ and, for $r > r'$, $\dfrac{p(r)}{p(r')} > k$, where k is big enough,

then an optimal auction can be described by Table 9.1.

Examining the above "Notes on Auctions for Pollution Reduction" by Dr. Maskin, the setting to achieve the desired goals was accomplished by establishing the creation of incentives for firms not to do "what's ethically wrong" and establish a "social benefit." Even though we can display economic advantages of achieving the desired goal, sustainable development entails "the development that meets the needs of the present without compromising the ability of future generations to meet their own needs" (Brundtland Report). Sustainable development comprises an ethical element and principals as the preservation of wildness and the radical rejection of an ethic of human use in favor of an ethic of human respect and non-interference of natural systems.

TABLE 9.1

Optimal Auction (Eric Maskin, Personal Communication, March 5, 2018)

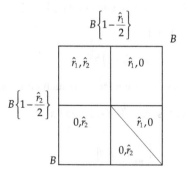

In this case, I refer to "mechanisms" as the adoption of strategies when an interaction between nations, cities, communities, and people is needed to achieve specific goals, with social connotations benefiting society as a whole and the people who are active participants in the process. In consequence, a mechanism needs to be designed and, in this case, the goal is to reduce air pollution and minimize health risks.

Human health has always been influenced by climate and weather, and severe events such as heat waves, droughts, wildfires, cold waves, snowfall, and flooding can all affect air and water quality, and consequently exacerbate the health risks to vulnerable populations. By definition, health is the state of complete physical, mental, and social well-being and not merely an absence of disease or infirmity (McMichael, 2003). This definition can help us to understand health risks and have a clearer picture of what we are trying to accomplish considering the inverse definition would be "the presence of a disease and lack of completeness in physical, mental, and social/emotional well-being tantamount to health risk" (Omoruyi and Kunle, 2011). The international frameworks that incorporate best practices for organizations, ISO 31000:2009 (risk management) and ISO 22301:2012 (business continuity management), define risk as the "effect of uncertainty on goals." This effect, which represents a deviation from the expected, can be positive or negative. Risk is usually characterized by events and the consequences of those events. Uncertainty is a lack of information that makes it difficult to understand the event, its consequences, or its likelihood (Pojasek, 2013; Marolla, 2017). Health risk is the population's susceptibility and degree of exposure to hazards and can be described as "chance of loss and/or variability from health and worsening of ill health" (Omoruyi and Kunle, 2011; Marolla, 2017).

Air pollution is a major environmental risk to health. By reducing air pollution levels, countries can reduce the burden of disease from stroke, heart disease, lung cancer, and both chronic and acute respiratory diseases, including asthma. In the case of firms emitting polluting GHGs as presented by Professor Maskin, the ethical and health risks aspect of the strategic approach is justified. Consequently, the mechanism design strategic approach addresses "social preferences such as altruism, reciprocity, intrinsic motivation, and a desire to uphold ethical norms which are essential to good government, often facilitating socially desirable allocations that would be unattainable by incentives that appeal solely to self-interest" (Bowles et al., 2008). Moreover, while promoting economic growth, mechanisms support a concise framework of addressing issues such as public health risks (Bowles et al., 2008) and supporting SDGs.

In this context, nations battling to achieve SDGs need to seek the economic, environmental, and social factors of the problems they face to sustain a balance and agreement in the approach to find a solution. A mechanism designer is a person/leader/institution/decision-maker who has to find a way to make several people/entities/nations/agents interact and select the rules with a concrete goal to achieve; therefore, in the case of countries working to achieve SDGs, they must agree in predetermined mechanisms to "play the game." Mathematically speaking, they need to define a game between several agents, where the nation leaders are the mechanism designers of strategizing the mechanisms to reach emission reductions. The issues nations face to implement SDGs today is that leaders do not behave as they are expected. Special interests and other factors such as the current economic and social situation, political environment, etc., are influencers of the outcome. Henceforth, as mechanism design proposes, they first need to define their objectives (goals) and work in reverse as it starts at the end of the game, then goes *backwards* (it is also called *reverse* game theory) (Bharathi et al., 2008). It is up to each country to determine its institutional arrangements for promoting, coordinating, and reviewing the implementation of SDGs and there is no single model. Because SDGs are complex in nature, mechanism design applications are optimal to consider methods on how to design and apply good system-wide solutions to problems involving several self-interested countries trying to promote their own agendas. Although SDGs are universal, each country will do what's best for them. Therefore, mechanisms will help to design institutions to solve resource allocation problems creating markets to efficiently allocate multiple heterogeneous objects, which is crucial to successful policy implementation. In this case, we must consider the cost borne by nations for the reduction of air pollution they have created. The challenges ahead are government and market failure to attain the desired outcome.

Government and Market Failure

Market failure occurs when a problem that violates one of the assumptions of the first welfare theorem and causes the market economy to deliver an outcome that does not maximize efficiency. Externalities arise whenever the actions of one economic agent directly affect another economic agent outside the market mechanism (Stantcheva, 2017). As presented in this chapter, some countries/institutions/firms have an incentive to hide relevant information. Use markets to correct market failure. It doesn't need to be costly and counter-productive. We need to consider a new paradigm shift in government and markets, supporting the development of new methods and moving the incentives to where the information is to motivate rational actors to behave in socially desirable ways. Nations' leaders must think strategically and have a solid framework or agenda as guidance to efficiently work toward those goals. Mechanism designers are institutions or people, choosing the right rules of their mechanisms to determine their objectives accordingly.

The challenge in achieving SDGs is that we usually consider the government to be the mechanism designer and that cannot be the case. All institutions, private and public sectors, must participate in some level of commitment. Manifestly, the leaders in the government sector might have their own interests like staying in power, policy restrictions, and obstacles such as policy goals inconsistent with the triple bottom line of sustainability, lack of policy communications, and knowledge exchange. Then, the society has to "design" a designer. The United Nations can be seen as the global "designer" to achieve SDGs. However, it is very difficult to monitor countries and countries' leadership initiatives because of the complexity of countries socio-economic and policy status and their own special interests.

Another example is the energy markets, ensuring access to affordable, reliable, sustainable, and modern energy, as highlighted in Sustainable Development Goal 7. Every company has its energy cost function and they need to decide the price distribution in order to maximize its expected total profit (Tiansong et al., 2012). The incentives of the companies are to announce a supply function that provides increasing prices, resulting in more profits. The expectation is that not all companies will be truthful. Mechanisms provide incentives to promote truth-revelation from agents, such that an optimal solution can be computed to the distributed optimization problem (Classic Mechanism Design, 2010). Mechanisms designed to foster coordination and collaboration among countries and develop a pragmatic strategy to cope with risks and build the foundation for economic property, social equality, and environmental justice are essential in the development of SDG strategy. Hence, its implementation helps to achieve overall development plans, reducing future economic, environmental, and social costs, strengthening economic competitiveness, and reducing poverty (Sustainable consumption production, 2016).

Risk Mechanism Theory

Socio-economic and environmental conditions are an important determinant of the likelihood that individuals and populations are exposed to risk factors. It is necessary to characterize and quantify the magnitude of risk for specific highly exposed, highly sensitive, or highly susceptible subgroups within the larger population. I call it Risk Mechanism Theory (RMT), as the unified methodology for creating mechanisms to attain a desired outcome (solution) addressing risk factors that affect societal balance. Henceforth, RMT represents a methodology to assess and proactively take action addressing social, economic, and environmental vulnerability conditions by measuring exposure to risks, assessing the intensity rate of impacts, and resilience level capacity according to the nation's adaptability, susceptibility, and resilience toward detrimental impacts affecting their stability and well-being.

This mechanism is a tool that provides an amalgamated and comparative measure of the risks of a nation's current situation. To provide a comprehensive analysis of a country's SDGs progress and achievability, it is important to understand its capacity of assessing risks and resilience. Furthermore, its vulnerability and exposure to hazardous conditions destabilizing society and the economic strength of the nation build a case to develop a risk mechanism framework parallel to the SDG agenda.

The lack of resilience and efficient economic growth creating the nations' inability to cope with risks and leading to wealth-distribution inequality jeopardizes sustainable development and creates a society of instability affecting good governance. Therefore, to measure a nation's capacity and vulnerability to achieve sustainable development we use the risk mechanism theory index (RMTI) that is composed of three elements described as follow:

- RMT Risk Exposure (RMT_{re})
- RMT Social and Economic and Environmental Vulnerability (RMT_{seen})
- RMT Resilience Level (RMT_{rl})

Henceforth, to calculate the risk exposure and vulnerabilities related to the population residing in the affected area the risk mechanism theory index formula displays the following partitions:

$$RMT = \left(\frac{RMT_{re} \times RMT_{seen} \times RMT_{rl}}{P_d} \right) \geq 0$$

Hence, RMT is the product of multiplying RMT risk exposure (RMT_{re}), social-economic and environmental vulnerabilities (RMT_{seen}) and resilience level capacity (RMT_{rl}) divided by the density of the population affected by the risk and impact event.

Population density (P_d) is one of the key parameters for assessing the magnitude of population exposed to risk, and the better quality data we have regarding the population susceptibilities to risks – population growth rate, economic conditions, levels of poverty, infrastructure and communication conditions, technology applications and developments, and disadvantage demographic communities – the better the assessment of risk.

Risk mechanism theory index will allow nations, cities, and communities to rank risks according to their probability of occurring divided by the population density, resulting from a specific activity or event.

Economic, social, and environmental vulnerability index helps to assess the capability of a nation to face risks and indicate the lack of access to resources, information, and technology and a vulnerable population's level of risk. All the factors and consequences of risks impacts are greater or equal to 0 (no impact).

The following composite risk mechanism theory mathematical expression assesses and identifies in more detail a set of core drivers with a significant impact on a nation's exposure and resilience to disruptive events. The expression or set of values below (Risk Mechanism Theory Index: RMTI) can be evaluated to produce a specific answer, as we have gathered the necessary data and assessed the risks to determine the value of each term:

$$\text{RMTI} = \frac{\sum_{c=seen}^{t} \sum_{\gamma_i=1} \widehat{w_{i(a,b,..)}}^{\,c}\ i^{*}_{r*[re*rl]}}{p_d} \tag{a}$$

Where C represents the indicator category for the social (S), economic (E) and environmental (E_n) vulnerabilities (*seen*); γ_i represents the indicator and t represents the total number of indicators within the respective category (C). $\widehat{w_i}$ represents the weighting evidence factor utilized for each indicator.

i^{*}_r is the intensity rate of the impact (social, economic, and/or environmental), where the multiplication (.) of the aforementioned factors (*re*, *seen* and *rl*) will provide the total weight intensity of the risk impact experienced.

Measuring i^{*}_r takes into consideration several factors affecting the situation/place. These conditions can be named as follow:

1. Severe weather events;
2. Cyber-attacks;
3. Environmental pollution;
4. Economic impact of disruptions/market disruptions;
5. Political instability;
6. Governance ineffectiveness;
7. Ineffective regulatory quality/deficient private sector development;
8. Unbalanced arbitrary exercise of power/Rule of Law;
9. Levels of corruption in government/public and private entities;

The following tables display the economic, environmental, and social impacts intensity (i^{*}_r). It is important to highlight the intensity rate is in correlation to the type of risk impact that will have different levels of socio-economic and environmental consequences for the country. The likelihood and impact assessment can be obtained from the following (Table 9.2):

Low: The consequences of the risk are minor, and it is unlikely to occur.

Moderate: Somewhat likely to occur, these risks come with slightly more serious consequences. If possible, take steps to prevent medium risks from occurring, but remember that they are not high-priority and should not significantly affect organization or project success.

High: These are serious risks that both have significant consequences, and are likely to occur. Prioritize and respond to these risks in the near term.

Extreme: Catastrophic risks that have severe consequences and are highly likely to occur. Extreme risks are the highest priority. You should respond to them immediately, as they can threaten the success of the organization or project (Table 9.3).

The weight of evidence table identifies the degree of the risk and impact weight in order of intensity and supporting RMT strategic approaches to a wide range of proposed remedies, including changes in regulatory statutes, policies, adaptation and the development of new methods for assessing risk.

The risk mechanism theory risk exposure, socio-economic, environmental, and resilience capacity indicators ($RMT_{re, seen, rl}$) provide an overview of different risks a nation can be affected with that hinder the SDGs realization underlying

TABLE 9.2

Likelihood and Impact/Intensity Rate Measurements

Likelihood	Impact				
	Insignificant	Minor	Moderate	Major	Severe
Almost certain	Moderate	High	High	Extreme	Extreme
Likely	Moderate	Moderate	High	High	Extreme
Possible	Low	Moderate	Moderate	High	Extreme
Unlikely	Low	Moderate	Moderate	Moderate	High
Rare	Low	Low	Moderate	Moderate	High

Source: Adapted from Risks Assessment Matrix (2018).

TABLE 9.3

Weight of Evidence and Category

Weight of Evidence	Category
1. Sufficient evidence from empirical studies and quantitative and qualitative data analysis to support a contributory relationship between risk exposure to the agent and the consequences of the intensity of the impact	$\widehat{w_{ia}}$
2. Limited evidence from empirical studies and sufficient data	$\widehat{w_{ib}}$
3. Sufficient evidence from empirical and quantitative and qualitative data studies but inadequate or no evidence or no data from the aforementioned studies	$\widehat{w_{ic}}$
4. Limited evidence from empirical and quantitative and qualitative data studies or an absence of evidence or data	$\widehat{w_{id}}$
5. Inadequate evidence from empirical and data studies or no data	$\widehat{w_{ie}}$
6. No evidence from empirical studies in at least two adequate empirical studies, coupled with no evidence from quantitative and qualitative data analysis	$\widehat{w_{if}}$

Source: Marolla (2017).

the three dimensions of risks: risk exposure, socio-economic and environmental vulnerabilities, and resilience level capacity. Nations, in relation to their capacity to cope with disruptive events, are very vulnerable to risks. We are experiencing a rapid growth in complexity of global supply chains, enabled by their dependence on rapidly developing information technology (IT), which has created a network of information flows that is increasingly vulnerable to external attacks (cyber-attacks). Given that the physical flow of goods and services in supply chains depends on underlying information flows, supply chains are exposed to a greater risk of operational disruption from these cyber-attacks (cyber-risk) affecting the entire nation's function of governance (MIT Center for Transportation and Logistics, 2017). Furthermore, these disruptions could yield to a drastic loss in productivity, competitive advantage, and profitability that would most probably lead to economic failure and impoverishment if not managed appropriately (Mensah et al., 2015).

Table 9.4 shows some of the impacts under the three dimensions. Some of those impacts crossover the different dimensions because those detrimental impacts have the potential of influencing the whole spectrum of society in a variety of ways (e.g., a country's risk to exposure of severe weather events impacts public health and overwhelms the health-care system; it also can create social and economic instability).

Risk mechanism theory index (RMTI) is a methodology to strategize against adverse situations allowing individuals and communities to adapt to extreme circumstances and lessen their impact through strategy indexes that track the performance of an algorithmic assessment strategy performed in a prescribed sequence to achieve a goal. The RMT mathematical expression describes a process with a finite number of steps that will achieve a particular result we are seeking; in this case to deter and/or minimize risks toward the achievement of SDGs. The methodology clearly and transparently specifies all the actions that must be taken through processes with a mathematical expression that might cover a given strategy's decision, or share its mitigation strategy. Moreover, composite indexes are a useful tool to measure complex, multi-dimensional concepts that cannot be observed directly. In conclusion, RMT is a well-defined strategic approach to global risks and serves as a comprehensive tool to address and find solutions to achieve SDGs.

Measuring risks is an important component of the RMT methodology. The mathematical equation (a) supports a concise method of evaluating risks and allows for the methodology to apply the most feasible adaptation and mitigation process. Financial disaster preparedness is vital and needs to be included with the RMT strategy. Therefore, a business continuity management system using ISO 22301 must be integrated with RMT. Physical and emotional recovery is a long process affecting individuals, family members, and entire communities, even countries. Financial disaster management assistance needs to be in place; however, additional problems affecting financial recovery often occur because key areas of risk were not adequately assessed, or their potential impacts were not fully understood. RMT aligns the structural

TABLE 9.4

RMT Three Dimensions Indicators Accordingly to the Risk Type Impact: Risk Exposure, Socio-Economic and Environmental Vulnerabilities and Resilience Level Capacity

	RMT	
$RMTI_{re}$	$RMTI_{seen}$	$RMTI_{H}$
Risk exposure to:	Socio-economic and environmental vulnerabilities:	Resilience level capacity:
Natural and man-made disasters	Fiscal stability/Productivity	Risk management strategy
Climate change	Inequality/wealth distribution	Business continuity systems
Inherent cyber-risk	Energy consumption and supply	Natural hazard risk quality
Health hazards	Control of corruption	ICT infrastructure and systems implementation
Scarcity of food and water	Human rights	ICT supporting electric grid standards
Disruptions in energy production	Environmental pollution	Energy infrastructure
Obsolete ICT systems	Governance, participation, good administration, access to justice, media and ethics	Urban infrastructure
ICT lack of knowledge dissemination/Digital divide	Rule of Law	Societal interdependencies
Environmental pollution	Political stability	Health risks
Biodiversity, flora, fauna and landscapes	Urbanization rate	Hospital capacity
Pandemics	Climate change	Health care
Accelerated urbanism	Environmental pollution	Accelerated urbanism
Population growth	Pandemics	Population growth
Inflation rates	Accelerated urbanism	Governance
Food supply and water	Population growth	Adaptation to disruptions in telecommunications, infrastructure
Vicious cycle of resource degradation – over extraction – degradation	Inflation rates	Recycling, demand rationing,
Isolation, semi-closeness, poor mobility, high cost of: mobility, infrastructural logistics, support systems, and production/exchange activities	Food supply and water	Renewable energy sources plan in place
Limited access to, and dependability of, external support (products, inputs, resources, experiences)	Physical degradation of land resources	ICT transportation management system
Environmental and health risks of socio-economic inequalities	Social inclusion and protection of particular groups	Growth of new industries technology driven flood management infrastructure
	Poverty indices; distribution and proportion of income dependent on risky resources; dependency and stability.	Water and food security
	GDP per capita; relative inequality; qualitative indicators of institutional arrangements.	

Source: Marolla (2017).

and non-structural adaptation and mitigation framework that build resilience and a well-rounded preparedness process to cope with risks. Severe economic downturns occur periodically and have grave consequences not only for the nation *en bloc*, it does impact the poor in a greater effect.

RMT elaborates a series of risk inputs to conduct an all-inclusive assessment of the economic risks faced by a nation. The following categories identify risks inputs that need to be included in the developing of the RMT Index (RMTI):

1. *Labor force*: Demographic changes, wars, and epidemics such as HIV/AIDS shock may lead to a large reduction in the quantity and quality (capability) of the labor force.

2. *Human capital*: Human capital risk may come from the lack of education, training, and experience.

3. *Physical capital*: All physical capital including the equipment, buildings, roads, and other infrastructure, information, and communication systems are liable to break down. The "breakdown" may only be partial or a total and sudden cessation of output production. Some breakdowns can bring a large part of the operation to a halt. Other failures might only have a significant impact if they occur at the same time as other failures. The sources include wars, hurricanes, earthquakes, volcanoes, fires, droughts, and floods.*

4. *Production process risks*: This may include a break down in the chain of the production process, or in the division of labor across regions/countries.

5. *Productivity or technology shocks*: Technological progress has led to structural adjustments and transformation: positive for some but negative for others.

6. *Financial risks*: Giving its functions of channeling savings and allocating resources, financial systems are vital parts of the production and economic process. Due to their inherent nature such as information asymmetry and market imperfection, financial systems have many risks which may lead to systemic crises and breakdowns.

7. *Macroeconomic policy risks*: this may include those caused by imprudent fiscal and monetary policies and mismanagement of the economy, which put growth at risk.

8. *Institutional and political risks*: Broadly, this may include institutional arrangements in a country, such as incentive mechanisms, corporate governance, national governance, and the Rule of Law, level of corruption, and changes in political systems that could put growth at risk.

9. *Global and external risks*: Changes or shocks in global economic and natural environment which are beyond the controls of any one national government, or human being (Wan and Yao, 2001).

In summary, the financial assessment integrated into the RMTI needs to be built from a complex set of interactions between the various risk inputs factors described above.

The 2007–2009 financial crises presented a series of lessons that need to be learned within the risk management framework of deterring economic shocks. Failures in the management of systemic risk events deter further actions toward SDGs. These economic shocks influence public and private organizations and leaders must take a look into the key issues that deter the prevention and/or business continuity of the system such as weak stress testing and scenario analysis, which can be either insufficiently extreme, or which fail to consider correlations between certain risk types (e.g., market and regulatory pressures, group dynamics, culture, risk perception, personal agendas and incentives, etc.) and their technical systems and processes (for example their operational systems, governance processes, risk management, reporting frameworks, etc.) (Simon Ashby, 2010). RMT is in the position to deliver a systematic strategy to cope with financial risks designed to reduce the probability that the assumed risks will in fact take place.

Information and Communication Technology's Crucial Role Driving Sustainable Growth and Social Inclusion Worldwide

Conservation of biodiversity involves dealing with problems caused by humans. The strategic planning supporting natural systems and restoring human damage needs to address the understanding of human attitudes, activities, knowledge, and behavior which are crucial to conservation research and build the foundation of an all-inclusive strategy.

Some of the main threats to our biodiversity are:

- Degradation, fragmentation, and loss of habitat;
- Spreading of invasive species;
- Unsustainable use of natural resources;
- Change of climate;
- Inappropriate fire regimes;
- Changes within the aquatic environment and water flows.

The mechanism to cope with the above risks needs to tailor human values, perceptions, judgments, and knowledge into conservation prioritization and decision-making. The strategic approach needs to improve and encourage greater use of these mechanisms in conservation science where the mechanism has to be designed to conform to some set standard. Therefore, the strategic planners need to design a "mechanism" by which a set of agents with productive capacities will interact with one another to allocate resources. In this case, the principal or planner needs to design a mechanism of interaction among the economic agents such that an appropriate efficient allocation of resources is achieved.

The policy-maker, for example, faces a problem of aggregating the announced preference of multiple agents into a collective or social decision. In regard to sustainable development, the goals are tailored on a local level, but they are global in context. Communication between agents is vital to develop the proper actions towards the problem faced. ICTs contribute to support the combination of individuals, groups, or countries interacting with each other to benefit the whole community. Hence, economic networks use the various competitive advantages and resources of each member to increase the production and wealth of all the members. In that context, ICT tools can enhance the interconnection of communities to collaborate and support existing mechanisms and systems to cope with environmental risks (Narahari, 2003).

Poverty is a complex issue with different causes and effects: lack of basic resources, access to information that is vital to their lives, and source of revenues as well as information about market prices for the goods they produce, about health, about the structure and services of public institutions, and about their rights. They have also limitations in political participation and influence in the institutions and power relations that shape their lives and the community they live. The inaccessibility of education makes the poor unable to develop the necessary skills to improve their livelihoods and participate in community activities for their well-being. They often lack access to markets and institutions, as both governmental and societal factors restrict them from accessing needed resources and services (Greenberg, 2005).

The Fundamental Role of Information and Communication Technologies in Modern Economic Growth, Innovation, and Development

Economic development and entrepreneurship are vital for a path to sustainable development because it generates an important addition to the overall gross domestic product (GDP), particularly in developing countries. This is also the case for wireless technology and ICT infrastructure development. In many emerging nations, it is a major challenge to gain access to capital and market information because of the lack of functioning infrastructure and constraints on financial resources. Sustainable development challenges, from climate adaptation to food security and health, require new, efficient technological solutions. Therefore, the link between innovation and entrepreneurship are increasingly seen as essential components for economic and social prosperity. There are several important initiatives fostering learning, cross-fertilization, and codification of knowledge to support innovation policy. The innovation, technology, and entrepreneurship (ITE) global practice, for example, brings together innovation and entrepreneurship specialists across the World Bank Group in knowledge exchange forums and other learning activities. Moreover, The Innovation Policy Platform (IPP), developed by the World Bank Group and the Organization for Economic Co-operation and Development (OECD), is a web-based interactive space that provides easy access to knowledge, learning resources, indicators, and communities of practice on the design, implementation, and evaluation of innovation poli-cies (The Innovation Policy Reform, 2013).

Venture capital (VC) investments are generally perceived as high risk and high reward. In general, venture capital investments typically have high rates of failure with just a few investments earning high returns. It states fur-ther that market practice assesses a fund's performance on a portfolio basis against its peers of the same vintage, not on a stand-alone basis. The positive effects on private sector development of International Finance Corporation (IFC) investments in venture capital funds confirm that this type of financ-ing can be an important mechanism for fostering innovation, entrepreneur-ship, and growth of private enterprises. The report states that Multilateral Investment Guarantee Agency (MIGA) support helped jump-start private sector foreign direct investment in post-conflict situations in Mozambique and Nicaragua. The report concludes that the effectiveness of MIGA's inter-ventions to support innovation and entrepreneurship will be enhanced with improvements in the quality of its front-end work in assessment, underwrit-ing, and monitoring (Independent Evaluation Group, 2013).

Private capital is likely to be particularly important in achieving the SDGs, advancing policy recommendations on international trade and investment, economic policy, climate change, measurement and indicators, and natural resources management. The OECD presents the following significant points of the power of investment to accelerate economic development:

- Private investment is a powerful development enabler: delivered in the right way it can create jobs, build skills, spur innovation, pro-vide essential infrastructure and services, boost economies and strengthen standards in public and corporate governance.

- Investment, both foreign and domestic, needs to be scaled up signifi-cantly in the coming years to contribute to the SDGs agenda.

- More investment is not enough, however. It must also be of good quality. Even though private finance accounts for the lion's share of capital inflows to developing countries, its contribution to develop-ment is still to fully materialize. The SDGs explicitly call for quality investment to support this transformation.

- Tapping the sustainable development potential of investment means increasing the capacity of the local economy and the public sector, reforming framework conditions to make countries attractive invest-ment destinations, and promoting responsible business conduct along the length of global supply chains (Investment for Sustainable Development, 2015).

List of Terms

Bayesian Nash equilibrium: A strategy profile and beliefs specified for each player about the types of the other players that maximizes the expected payoff for each player given their beliefs about the other players' types and given the strategies played by the other players.

Continuous distribution: Describes the probabilities of the possible values of a continuous random variable. A *continuous* random variable is a random variable with a set of possible values (known as the range) that is infinite and uncountable.

Environmental pollution: *Defined* as "the contamination of the physical and biological components of the earth/atmosphere system to such an extent that normal *environmental* processes are adversely affected.

Ex ante: Based on assumption and prediction and being essentially subjective and estimative.

Ex post: Based on knowledge and retrospection and being essentially objective and factual.

First fundamental theorem of welfare economics (also known as the "Invisible Hand Theorem"): Any competitive equilibrium leads to a Pareto efficient allocation of resources. The main idea here is that markets lead to social optimum. Thus, no intervention of the government is required, and it should adopt only "laissez faire" policies. However, those who support government intervention say that the assumptions needed in order for this theorem to work, are rarely seen in real life.

Incentive-compatible (IC): When every participant can achieve the best outcome to him/herself just by acting according to his/her true preferences. There are several different degrees of incentive-compatibility: The stronger degree is dominant-strategy incentive-compatibility.

Lagrange Multiplier: A strategy for finding the local maxima and minima of a function subject to equality constraints. The great advantage of this method is that it allows the optimization to be solved without explicit parameterization in terms of the constraints

Mechanism design: A field in economics and game theory that takes an engineering approach to designing economic mechanisms or incentives, toward desired objectives, in strategic settings, where players act rationally.

Private capital: Money provided to a business as a loan or *equity* investment that does not come from an institutional source, such as a bank or government entity, or from the public through selling stock on a stock exchange.

Private investment: Money invested by companies, financial organizations, or other *investors*, rather than by a government.

Policy-maker: A member of a government department, legislature, or other organization who is responsible for making new rules, laws, etc

Probability distribution: A function that describes the likelihood of obtaining the possible values that a random variable can assume. In other words, the values of the variable vary based on the underlying *probability distribution*.

Revelation principle: A fundamental principle in mechanism design. It states that if a social choice function can be implemented by an arbitrary mechanism, then the same function can be implemented by an incentive-compatible-direct-mechanism with the same equilibrium outcome.

Risk Mechanism Theory (RMT): Unified methodology for creating mechanisms to attain a desired outcome (solution) addressing risks factors that affect societal balance.

Risk Mechanism Theory Index (RMTI): A risk assessment methodology to address global multi-complex issues presenting a well-defined framework's problem of achieving some objective or maximizing some utlility function subject to the level and category of the risk.

Social benefits: Private *benefits* gained by individuals directly involved in a transaction together with the external *benefits* gained by third parties not directly involved in the transaction.

Strategy: The art and science of planning and marshalling resources for their most efficient and effective use.

10

Innovation Strategy

The Art and Science of Executing Your Vision

As a general rule, organizations in the private sector emphasize the need for growth through a series of initiatives and strategies to create "wealth." This entails the benefits of having a growing business and providing the foundations to workers' well-being, the community, and in some cases, the society as a whole. The twenty-first-century challenges are obviously greater than just achieving "revenue growth." It is vital to engage in a series of strategies to benefit society as a whole and cope with risks and changing technologies that are fundamental to an organization's future success in every aspect of its operations and impacts. Organizations must unleash the three dimensions of sustainability: environmental, economic, and social, which are all interlinked. The social impact of a business is easy to identify but difficult to measure; however, understanding the effects a company has on society and the environment is vital to achieving sustainability.

Harvard professor Clay Christensen coined the term disruptive technology, which he later renamed disruptive innovation. He identified that it was not the technology itself which was disruptive, but its impact on strategy or business models. Christensen's theory developed the previous body of literature about the discontinuity of organizational change. In summary:

1. A disruptive technology emerges. Initially, it cannot match the performance of the existing dominant technology, on the factors which customers traditionally value.

2. The distinctive features of the disruptive technology are valued by a small fringe segment of customers and increasing numbers of new customers. It is also typically cheaper, simpler, smaller, or more convenient. Incumbent players in the market conclude that investment in the disruptive technology is irrational, since their most profitable customers don't want and can't envisage using new products based on the new technology.

3. New entrants to the market who exploit the disruptive technology concentrate on fringe or emerging markets.

4. The disruptive innovation develops so that new products meet the standards of performance expected by the bulk of the market; the new technology displaces the previous one and thus the new entrants to the market displace the incumbents.

Leading Change in the Digital Age

I strongly believe citizens' participation is fundamental in achieving the ambitious sustainable development targets. Also, they will depend on a number of other factors, political will and adequate funding among them. However, without strategy, the goals will never be achieved. The strategic regime must be applied throughout the system of communities, cities, and nations from domestic initiatives to global frameworks for policy coherence. Change is necessary if the existing approach is not feasible to attain sustainable development goals (SDGs). Because "the concept of sustainable development implies, most importantly, the consideration of spatial (i.e., "where") and temporal (i.e., "when") dimensions, system components (i.e., economic, social and environmental aspects) and, in particular, the interactions between these components (i.e., "why, who, how, how much"), it is necessary to understand the synergies to have an adaptive and flexible approach to solve SDGs obstacles to success. As sustainable development refers at once to a process (development) and to a condition (sustainable), development, like all system processes, is dynamic and a function of its dimensions and components (Winograd and Farrow, 2011).

That is why strategic flexibility and adaptability are highly important to obtain concrete results that are the place of mechanism design, risk management frameworks and strategist adaptive cognition (SAC). Leaders cannot take a one-size-fits-all approach to change. They must adapt and change accordingly to the challenges and the feasibility of materializing the goals with the proper strategy. If adaptive leadership is not taking place the strategic planning will be jeopardized and most likely it will face resistance. Professor Kotter presents a comprehensive analysis of strategy and management. It is important to notate that strategies can work only when participants have the information they need to provide useful input because it will be inefficient if they don't. Moreover, in many cases involving private and public entities and nations, leaders often don't tailor the speed of their change strategy to the situation. For instance, they may apply a go-slow approach even when an impending crisis calls for rapid change (Kotter and Schlesinger, 2013).

Powerful Guiding Coalitions with a Comprehensive Strategy

Meaningful changes in any organization are full of obstacles and in regards of nations leading change to meet SDGs, the entrenched culture can be too strong that could potentially deter them from making the necessary changes and developing a shared commitment to achieving the necessary objectives. What does it take to transform the status quo and deliver the desired results? As Harvard professor John Kotter stated:

> individuals alone, no matter how competent or charismatic, never have all the assets needed to overcome tradition and inertia except in very small organizations. Weak committees are usually even more impotent. Efforts that lack a powerful enough guiding coalition can make apparent progress for a while. The structure might be changed, or a re-engineering effort might be launched but sooner or later, countervailing forces undermine the initiatives.

> **Kotter, 1996**

Therefore, a comprehensive strategy that involves the identification of risks and assessing whether the strategic planning can successfully implement changes that confront them must be enforced with the mobilization of financial resources, capacity building, and technology, as well as data and institutions. Nevertheless, I highly recommend an economic and social impact assessment of the development projects in consideration with the strategy, highlighting the implementation as high, medium and low risk.

The Financial Aspect of Implementing the Strategy

The local implementation of the strategy is crucial for effectiveness. Although the goals and purpose of the strategy are global, taking actions regionally can have a long-lasting effect. The partnership between the private and public sector, along with the need to strengthen domestic resource mobilization and management, must be the foundation of the strategy. Countries and regions in general have differentiated needs, priorities, and socio-economic vulnerabilities and capabilities. Therefore, developing domestic financial strategies for sustainable development is needed with systematic costing efforts to support projects that need to be prioritized. Along with the aforementioned strategic planning, local leaders must have in place policy coherence and coordination between different policy processes and institutions that will be critical for an efficient outcome. Although there are substantial variations between countries, any domestic and/or regional financial system has to be aligned with the sustainable development strategy. There is a need for increasing crossed-institutional collaborations and public and private capital that could potentially be tapped to help deliver the SDGs. National-level development finance institutions in each region can also play an important role in directing investment flows toward strategies to support the SDGs (Implementing the Sustainable Development Goals, 2017).

Underestimating the Power of Vision

It is clear that the world will never reach the SDGs without businesses. While businesses can make positive contributions, such as creating jobs, finding innovative solutions for climate challenges, or contributing to human capital development, they can also cause or contribute to negative impacts, such as exploiting labor in supply chains, damaging the environment, or engaging in dishonest practices. Businesses should focus on activities to ensure that they avoid undermining the SDGs by causing or contributing to negative impacts. Businesses need to look at its global impact and influence beyond compliance and risk management addressing the civil society positive development impacts that will be achieved through improved treatment of the millions of workers and communities affected by business activities around the world.

It is important to lay down the goals with a sense of urgency and a strong guiding team, but they are insufficient conditions for major change. Leaders must advocate a paradigm shift in the way they think about and do development. As Harvard professor John P. Kotter explained: "of the remaining elements that are always found in successful transformations, no one is more important than a sensible vision. Vision plays a key role in producing useful change by helping to direct, align, and inspire actions on the part of large numbers of people. Without an appropriate vision, a transformation effort can easily dissolve into a list of confusing, incompatible, and very time-consuming projects, which take you in the wrong direction or nowhere at all" (Kotter, 1996).

Strategic Adaptive Cognitive Framework

Strategic Disruption

The strategy in place and the leaders in charge need to create a strategic mindset and consequently build the plan and coalition to take lead in the SDGs. One of the aspects of such a strategy is the monitoring, learning, and adapting to the current situation and challenges to achieve the goals. The assessment of the economic, social, and environmental aspect of the local (for cities, rural areas) and national conditions has to tailor an integrated set of indicators to allow analysis of the inherent trade-offs and interconnections between the economic, social, and environmental dimensions of sustainable development. I would like to emphasize the importance of the way we do the analysis of the country's current situation and how feasible it is to implement tangible actions toward sustainable development, which are the main keys for leaders to achieve such objectives.

Strategic adaptive cognition (SAC) lays out the concept of creating a strategic mindset that takes a broad, long-range approach to problem-solving and decision-making. Since SDGs take a broad global approach, planning ahead as well as understanding the current situation involving analysis and pragmatic steps is required to ultimately accomplish the strategic objectives. It is important to consider a strategy of understanding and acceptance of different approaches to change while enhancing willingness and ability to change. SAC also helps to identify the impact of the leaders' decisions on the different configurations of governance, which is the structural aspect of this complex adaptive system. An accurate picture and performance requires holistic assessment of its underlying structures with the identification of key players such as stakeholders, issues, processes, and relationships that are interlinked with the efficient operations of the system. Governance practices go hand-in-hand with the strategy and the leaders' vision to materialize SDGs. The strategy needs to evolve in response to internal and external agents, adapting to changed circumstances and essentially learn, adapt, and self-organize.

John R. Wells defines strategic intelligence as the capacity to adapt to changing circumstances, as opposed to blindly continuing on a path when all the signals in your competitive environment suggest you need to change course (Strategic IQ: Creating smarter corporations, 2012). SAC's concept originates from problem-solving approaches and novel strategies that are often best learned analytically and experientially. There are seven steps:

- Systematic Thinking: Enhancing the ability of a person or organization to use learning effectively is crucial to comprehend the implications of any critical situation. Organizations and nations can learn from each other's failures and successes by continually expanding their capacity to create the results they need to accomplish. When new and expansive patterns of thinking are nurtured, where collective aspiration is set free, and where people are continually learning to see the whole together, goals become tangible and results will be materialized.

- Strategic Approach to Management and Policy: A forward-thinking strategy that addresses the three responsibilities of an organization: Social, Environmental and Financial. It needs the proper systems implementation to pursue sustainability and evaluate the impacts of sustainability on social, environmental, and financial performance along with the trade-offs that ultimately must be made, putting emphasis on the present, considering future outcomes, and long-term strategies to cope with climate change impacts.

- Adaptive and Flexible Behavior: Behavioral flexibility is employed by many species as an adaptive response to changing environments. Flexibility in behavior lies at one end of a continuum of plastic responses that includes developmental plasticity in individual physiology and anatomy, and genetic responses to selection over generations (Dukas, 1998b; Pigliucci, 2001; West-Eberhard, 2003). Of these forms of response, behavioral changes can generally occur most quickly and therefore are best suited for rapid responses to changes in the external environment. As climate change impacts place societies and populations onto a changing environment that exacerbates existing risks and increases vulnerabilities, the adaptation of leaders to environmental change affecting their organizations becomes a fundamental asset for understanding such threats and assimilating the necessary knowledge and information to find strategies and solutions to those problems. Strategic flexibility goes along with behavioral flexibility in adapting to external and internal changes and challenges that threaten the existing structures and/or systems that cannot deal with the new risks.

- Strategic Cognition: learning from other leaders, strategies, and experiences and subsequently applying them to evolve a strategy for improvement. Past experiences deliver the path for a concrete course of evidence and then take them into account to plan, do, check, and act accordingly to address climate change risks. This requires being people-centric. Leaders need to have the ability to accomplish goals and objectives by operating in cross-cultural contexts with interpersonal skills and adaptive thinking.

- Critical Thinking and Learning: Assimilation and understanding information can be biased, distorted, partial, uninformed, or downright prejudiced. Yet the quality of life and what we produce, make, or build depends precisely on the quality of our thought. For leaders, fostering critical learning and thinking habits becomes crucial to being prepared to confront important issues, much more so than to simply learn facts. "Habits of Mind are the characteristics of what intelligent people do when they are confronted with problems, the resolutions of which are not immediately apparent," (Costa and Kallick, 2008).

- Experiential Training: Develop behavioral skills and objective abilities. "Learning by doing" involves the interaction of participants with the goal of assimilating and understanding specific situations and applying them to their own reality, unlike the informational training methods which are more one-sided. Here the major focus is not just the mere transfer of facts and figures but also the development of skills in the participants, which may or not be the case in informational training.

- Shared Value and Logical Incrementalism: A change of mindset that requires the thinking of growth and the improvement of societal conditions. Shared value applies to businesses around the world and it can easily relate to the SDGs creating measurable benefits by identifying and addressing social problems that intersect with their organization implementing sustainability, social responsibility and sustainable development. Shared value, writes Initiative for a Competitive Inner City founder and Harvard Business School Professor Michael Porter, is defined as "policies and operating practices that enhance the competitiveness of a company while simultaneously advancing the economic and social conditions in the communities in which it operates… Shared value is not social responsibility, philanthropy, or even sustainability, but a new way to achieve economic success." When economic success is immersed in a system, sustainable development rises and poverty eradication can become materialized (although, not all economic progress is equal to poverty eradication). Logical incrementalism goes along with shared-value strategies as it states that strategies do not come into existence based on a one-time decision but rather, it exists through making small decisions that are evaluated periodically. These small decisions are not made randomly but logically through experimentation and learning. Embracing their inter-dependencies within communities and strategically including community impact in the strategy can produce measurable advantages and leverage private development money. It is important to note that the work of city leaders is primarily made up of mediation and collaboration and has the unique ability to leverage both public and private money to create shared value (Porter and Kramer, 2011).

A vital tool for leaders is learning to utilize "what you know." Cognitive learning requires repletion, summarizing meaning, guessing the meaning from context, and using imagery for memorization. Recognizing problems and making changes to correct them often presents substantial challenges (Suharno, 2010). Therefore, using background knowledge, gathering new information to clarify, and forward-thinking objectives is the path for a successful plan. The recognition of problems at an early stage is highly important as such problems will be magnified over time and more damage will occur to the organizations; consequently, the difficulty to solve them would be greater. This can be exemplified by the "inverse square law," which states that a specified physical quantity or intensity is inversely proportional to the square of the distance from the source of that physical quantity (Appalachian State University, 2015). Applying this concept to business and/or organizations leads to the understanding that an important issue that is not resolved promptly will multiply and become difficult to control and/or fix in the future. Therefore, a focus to resolving issues, determining and addressing risks of implementing the SDGs will need the capability to enact major strategic changes to resolve problems in a timely fashion; otherwise the cost of delay will exacerbate those risks. This approach requires strategic flexibility and is composed of three elements: maintaining attention, completing an assessment, and taking action (Shimizu and Hitt, 2004).

Strategic Adaptive Cognitive Planning: Defining the Issues and Creating Solutions

There are three fundamental steps for effective strategic planning:

1. Understanding where the organization is going with the plan outlining the goals.
2. Mapping the strategy from methodology to concept to development.
3. Knowing how the strategy will be implemented.

The leadership team can drastically reduce the obstacles to develop a concrete strategy and employ actions to deliver SDGs by proactively seeking ways to evolve, implement, and continuously improve and adapt the strategic management plan. The strategy can be "local," but needs to coordinate efforts with a national strategy to decentralize and support decisions and resources for local involvement. It also needs to be secured in broad community input and strategic partnerships.

The Three SAC Planning Imperatives:

- Micro-Strategic Phase "Inside Out": The micro-strategy approach is closely aligned with the way military strategists have been trained for hundreds of years.

    In military science, a plan is laid down answering three questions:

    - "What do we want?" – Resulting in outcomes
    - "What do we have?" – Resulting in assets
    - "What will we do?" – Resulting in actions (Logan and Fischer-Wright, 2009; Marolla, 2017).

        SAC's strategic steps focus on key issues instead of on the distracting information that can conceal clear thinking and a path to a successful strategic planning. Logan and Fischer-Wright (2009) presented a process of answering strategic questions that are important to be considered to find the foundation of the strategic mindset:

        a. Leaders ask: *What do we value?* Answering this question can require deep reflection and clear thinking.
        b. Leaders ask: *What do we want?* They work, with a deadline, until the answer is specific, making sure that the goals are specific extensions of their values. This discussion results in defining the outcomes.
        c. Leaders ask: *What do we have?* Here, they pool the collective wisdom of everyone involved to make a long list, resulting in their inventory of assets. As a habit, leaders and strategists add to their assets on a regular basis.

d. Leaders ask: *Do we have enough assets for the outcomes?*

e. If the answer is no, the leaders adjust the outcomes or develop an interim strategy to build the assets they need, or both.

f. If the answer is yes, they ask: *What will we do?* The result is a list of actions.

g. Leaders ask: *Will the actions accomplish the outcomes?*

h. If the answer is no, they add more actions, paying special attention to how assets can be translated into actions. We're often asked how this model accounts for competitive advantage and other aspects of applied economics. The answer is that if the people engaged in setting a strategy are knowledgeable of the competitive landscape, the answers they provide to the questions in steps 4 and 7 will take that knowledge into account (Logan and Fischer-Wright, 2009; Marolla, 2017).

- Mezzo-Strategic Phase "Creating the Conditions": This strategic approach combines the elements of micro and macro strategies creating the environment for a successful transition of the initiatives starting off in the micro strategic stage towards the macro strategic plan. It consists of creating a change facilitative infrastructure, ensuring all the appropriate resources are in place and connecting the strategic planning, operations, and people to deliver actions. In this phase, leaders need to identify resources to support the development of the strategy. They also need to advocate for the micro and macro-strategic phase, setting goals, and expectations and linking the Plan-Do-Check-Act four-step model for carrying out change which is recommended before the macro-strategic phase is initiated.

- Macro-Strategic Phase "Outside In": Taking into consideration the external factors affecting the execution of SDGs building a facilitative culture and a structure of ongoing strategizing levels and develops a clear vision for meso/micro-strategic phase that can be tailored to the macro-strategy on a local and global context. While launching the initial steps of the strategy, clear distinctions between what the strategy, implementation, and execution will do and what it will not do (or implement/initiate) is essential to efficiently deliver the plan. Stakeholder's engagement, the public and private sector, improvements in the strategies and familiarizing them and involving them with the micro and mezzo phases helps continues improvement of the process. Creating an appreciation for the regulatory environment that is essential for good governance and identifying how internal and external forces influence all levels of the micro, mezzo, and macro strategies' execution.

SAC planning is not just a set of rules and rigid steps leaders must follow; it is the organization's culture and philosophy where leaders apply "the strategy" first and then deliver that thought to a process. Making sure the three phases are followed by strategic flexibility and adapted to each SDG can deliver higher chances of amalgamating the entire plan *and* organizational structure under the same goals to be successful in implementing the strategic plan. All parties in the governance's system can contribute to a common denominator of strategizing a framework. The decisions and activities leaders undertake in order to implement the strategy into a tangible success is accompanied by the realization of the best possible results a strategy and its implementation will allow. SAC planning can help in providing a framework to sustainable development as the central guiding principle for the future development of nations and to identify interdependences between the governance's departments to assess risks and build strategies to develop scenarios concerning the future scope for action. Ultimately, we need to find the strategic adaptability that encompasses synergies to gain knowledge and address obstacles with concrete solutions. SDGs are vulnerable to socio-economic and environmental factors. A comprehensive and multi-dimensional, cross-sectoral strategic approach and the integration of existing planning processes can be supported by systemic frameworks because they help to develop objectives that are measurable and time-based. Therefore, mechanisms enable systems to structure themselves to accomplish the pursued objectives. Moreover, mechanisms help us see, understand, and communicate the importance of interdependency between players in comprehensible and meaningful ways. It is fundamental for the effectiveness of the strategy to utilize the implementation of a mix of policy initiatives, including more use of environmental and economic instruments such as mechanized design frameworks and risk management planning that can have policy implications.

Business Strategy and Sustainable Development Goals

The importance of assessing and meeting the targets of sustainable development has high relevance to the outcome and the strategy needs to be aligned with the reporting of those goals that are most relevant to the organizations' structure and vision. An organization's structure plays an important role in how practicable the organization's actions are to achieve the SDGs. The type of business that measures the impact of its initiatives to accomplish the SDGs also plays an important role because if we talk about the ICT sector, these businesses will focus on applications of new technologies that can support and deliver tangible actions towards the goals. Moreover, the institutions' emphasis on specific SDGs as well as their capacity and budgets ready to be executed and quantify the SDGs progress, can deliver appropriate results with the suitable strategy.

Strategic Adaptive Cognition for Sustainable Development

SAC relates to the concept of sustainable development operational because it considers its multi-dimensional index combining the information derived from a selection of relevant sustainability indicators belonging to economic, social, and environmental pillars. It reproduces the dynamics of these indicators over time and countries. The SAC modeling framework is a recursive-dynamic strategic forward-thinking approach used to integrate sustainable development into their strategic planning and create long-lasting transformative change and measure overall sustainability. Macro-Strategic Phase "Outside In" takes into consideration the external factors affecting the execution of SDGs, building a facilitative culture and a structure of ongoing strategizing levels and developing a clear vision capturing the sector and regional interactions and higher-order effects driven by background assumptions on relevant variables to depict future scenarios. This makes it possible to compare sustainability performances to achieve the SDGs – under alternative scenarios, across countries, and over time – and select the best strategic approach based on empirical analysis. The actionable plans and results to sustainable development among countries and organizations fluctuate and do not always reach the preferred outcome. A concrete and detail-oriented, strategic approach such as SAC and mechanized design provides the rationale for thinking about the dedicated policies supporting overall sustainability to help shape development programs and influence development policy; showing that social and environmental benefits may be greater than the correlated economic costs. The use of strategic frameworks for sustainable development is a unity of efforts that requires leaders to understand the impacts of their decisions and integrate the corresponding actions taken into the long-term strategic goals.

Aligning Sustainable Development Goals to the Vision

Strategy, Structure, and Culture

Strategically aligned organizations fulfill their purpose and tailor their goals to the economic, social, and environmental pillars and responsibilities to society. This is true for countries and cities as well because each one of them is a system of systems. This model applies to the dependency of the entity's resilience to its systems such as transportation, utilities, socio-economic situation, environmental risks, and public health infrastructure and communication networks. These systems are individually dependent and also rely on how interdependent those systems are between them (Marolla, 2017).

To successfully implement and obtain positive results, bringing into line the three sustainability pillars, the structural system, and operations must follow concise interdependent components that make up a strategically aligned organization. Dr. Barry Varcoe and Jonathan Trevor, from the University of Oxford, have a focus on Enterprise Alignment – how all elements of a business are arranged in such a way as to best support the fulfillment of its long-term purpose. That can also be shaped and applied to the countries' SDGs strategic approach. As leaders we should ask the following questions:

- How well does our strategy fulfill the SDGs?
- How well do our organizational capabilities support the delivery of the strategy?
- How well do our resources enable development of our required organizational capabilities?
- How well do our management systems drive the performance of our valuable resources? (Adapted from Trevor and Varcoe, 2017).

Leaders with a strategic plan in place will be able to work around obstacles and challenges that are impeding the SDGs to materialize. However, it is important to align SDGs with the interests of the organization and make it easy for the decision-makers and the implementation team to make the right decisions every day. Those decisions need to be based on the longer-term strategy to achieve SDGs rather than the more immediate demands of people and politics. The vision, strategy, and goals should all be in place tailored to a detailed communications plan guiding every stage of the transformation, from initial launch to sustaining and building on the improvement. A transformation's top-down communications start with a compelling, personally inspiring view of the future (Adapted from Steve Sakson and George Whitmore, "Communications strategy: A vital (but often overlooked) element in lean management transformations" (Sakson and Whitmore, 2013).

List of Terms

Biodiversity: The variability among living organisms from all sources including, inter alia, terrestrial, marine and other aquatic ecosystems, and the ecological complexes of which they are part; this includes diversity within species, between species, and of ecosystems.

Capacity building: The process of developing and strengthening the skills, instincts, abilities, processes, and resources that organizations and communities need to survive, adapt, and thrive in the fast-changing world.

Cognitive learning: A theory that defines learning as a behavioral change based on the acquisition of information about the environment.

Disruptive innovation: An innovation that creates a new market and value network and eventually disrupts an existing market and value network, displacing established market-leading firms, products, and alliances.

Disruptive technology: The one that displaces an established *technology* and shakes up the industry or a ground-breaking product that creates a completely new industry.

Private sector: The part of a country's economic system that is run by individuals and companies, rather than the government.

Public sector: The portion of an economic system that is controlled by national, state or provincial, and local governments.

Social inclusion: The process of improving the terms on which individuals and groups take part in society – improving the ability, opportunity, and dignity of those disadvantaged on the basis of their identity.

Strategic adaptability: A planned ability to react effectively when business and environmental factors change unexpectedly.

Strategic adaptive cognition (SAC): A strategic mindset that takes a broad, long-range approach to problem-solving and decision-making.

Strategic flexibility: The capability of an organization to respond to major changes that take place in its external environment by committing the resources necessary to respond to those changes.

Strategic planning: A systematic process of envisioning a desired future and translating this vision into broadly defined goals or objectives and a sequence of steps to achieve them. In contrast to long-term planning (which begins with the current status and lays down a path to meet estimated future needs); strategic planning begins with the desired end and works backward to the current status.

Venture capital (VC): Financing that investors provide to startup companies and small businesses that are believed to have long-term growth potential. Venture capital generally comes from well-off investors, investment banks, and any other financial institutions.

11

The Information and Communication Technology Revolution

Promoting Economic Growth, Combating Climate Change Impacts, Reducing Health Risks, and the Leader's Role in Strategy

Number 8 of the 17 proposed Sustainable Development Goals (SDGs) is "to promote sustained, inclusive and sustainable economic growth, full and productive employment and decent work for all" (Sustainable Development Goal 8, 2017). The economic prosperity of a nation and the environmental impact of the nation's growth have been studied extensively and the results are controversial. There is a reason we need to make a balance between a successful economy and the preservation of natural ecosystems. Therefore, economic growth alone is not enough to improve environmental quality. Sustainable Development Goal 17 (means of implementation) includes Target 17.14, to: "enhance policy coherence for sustainable development" (Policy Coherence for Sustainable Development in the SDG Framework, 2015). International policy accountability for policy coherence implementation is needed to further move the agenda and achieve the goals. Developing an effective environmental policy framework is essential in order to improve environmental quality that supports well-being and enables long-term economic development (Almeida et al., 2017). Inclusive economic growth, combating climate variability and change risks, infrastructure development, addressing population growth and urbanism, international trade, and international financial system concerns, are among some of the issues that raise the importance of redistribution as the most effective way to poverty eradication and sustainable development. Economic growth does not go hand-in-hand with economic development. To overcome the obstacles to accomplish SDGs, leaders must strengthen the capacity of governments to take into action the mutually supported policies that build upon inclusive strategies and technology innovation. Economic activity depends on the natural environment and so maintaining the condition of natural assets is a key factor to sustainable development and growth. Also, economic growth contributes to the investment and dynamism needed to develop and deploy new technology, which is crucial to both productivity growth and managing environmental assets. Environmental assets contribute to managing risks to economic and social activity. The managing of risks with the use of information and communication technologies (ICTs) must be an essential tool to policy-makers because a concrete implementation of a risk management framework helps to cope with severe weather events and its consequences such as flood risks, heat waves, regulating the local climate (both air quality and temperature), and maintaining the supply of clean water and other resources. This strengthens economic activity and maintaining the condition of natural assets (Everett et al., 2010).

Conclusion

Technology won't solve the biggest issues the world faces today. People do. Therefore, the future is now, because the digital revolution we are experiencing is not solely about technology but is also about people. The influence of ICT is reflected in everyday activities and in our increasingly tech-enabled world, leaders of the private and public organizations and government officials have a more crucial influence in shaping, deploying and powering new disruptive technologies that transform life, businesses, and the global economy. Leaders must be disruptive in order of having the potential to effectively change the status quo. Rethinking leadership for the new digital age is a priority to cope with existing risks and solve new ones. As technology advances in an accelerating phase, there is an urgent need for strong direction to lead in the digital transformation to address the economic, social and environmental challenges of the twenty-first century. I will argue that achieving SDGs through ICT requires a paradigm shift that reshapes not only the economy but also society at large, and even how we think and perceive our world. There are many private organizations and NGOs that are taking the lead to implement this digital transformation. However, advancing sustainable development with ICT as the driving force requires the implementation of the new technological revolution with a comprehensive strategy powerful enough to manage and interconnect nations and people to make structural changes supported by evidence-informed innovations and cognitive leadership. Two issues of profound significance lie at the heart of our current world affairs; the challenges of environmental and socio-economic vulnerabilities and the potential of ICT in promoting SDGs. Major empirical and methodological strategies that deal with the complexity, irreversibility, and uncertainty of the socio-economic and environmental challenges over the long term must be applied for achieving a better understanding of our environment and the natural ecosystems. There is a lack of concrete actions from leaders to build sustainable development initiatives to mobilize appropriate science and technology. Science-based knowledge and technological innovations tackling the SDGs are not placed in policy and programs adequately and that impede actionable planning for defining strategies able to deliver technological and innovation capacity, which is a cross-cutting issue for inclusive development and sustainable growth. As leadership is becoming more collective than individual, it is necessary to strategize locally with a global approach and influence. Comprehensive and pragmatic strategic planning with social and economic development in every aspect of policy-making is included in the proposed mechanisms and frameworks I presented in this book.

Market mechanisms are important means for supporting sustainable development. It supports the development and increase in the flow of investments, technology transfer and access to leading-edge clean technologies. We take into consideration that some markets are free of government intervention while others are regulated. By adopting ICTs with *sustainable* practices, companies can gain a competitive edge, increase their market share, and boost shareholder value. Henceforth, organizations addressing important issues targeting *sustainable development* goals from a *business* perspective support the efficient resource allocation needed to achieve sustainability. This way, organizations of the public and private sector work towards the international community's most urgent priority: sustainable development.

The strategic approach I presented using ICTs provides a coherent framework for analyzing this great variety of institutions, or "allocation mechanisms." Mechanism design focuses on problems associated with incentives and private information. Risk management frameworks deal with uncertainties or "risks" and argue for the need for an analytical framework that can support the hard pragmatic and scientific work to achieve the SDGs. Risk mechanism theory is designed for an assessment of socio-economic and environmental impacts. Strategic adaptive cognition (SAC) supports a comprehensive approach to an organization's strategy to custom-make and differentiate the characteristics of strategic leadership on a sustainable path. All above ensure that market and market mechanisms contribute to food security, environmental sustainability, poverty eradication, and economic development.

ICTs are catalysts for sustainable development, as the spread of ICTs and global interconnectedness has great potential to accelerate human progress, bridge the digital divide, and develop knowledge societies. This needs to be parallel to technological and scientific advancements that require developing and mobilizing research and people with public and private funding, and improved systems of global governance. Sustainable development mechanisms incorporate analytic considerations on climate, biodiversity preservation, and global poverty reduction, and provide critical strategies for consensus building to achieve the SDGs.

As indicated in the book's introduction: "Sustainable development entails the action of making effective decisions from some set of available alternatives for the best use of a situation or resource." Consequently, the strategic approach to SDGs is fundamental to delivering results. This means understanding how the potential *mechanisms* could assist in improving the quality and quantity dimensions of the projects aimed at SDGs such as sustainable economic growth, cost-effective programs, and the accessibility of those programs particularly to the poor, etc. The aforesaid are important elements of transformation.

There are several ways to approach the SDGs planning from development to implementation. Some initiatives and programs aim at emission reductions, policy and sectoral decision-making, technology, and governance. Nevertheless, the flexibility of mechanisms and strategic planning allows identifying the vulnerabilities of the strategy implementation phase and the inclusiveness of all strategies to materialize and work towards the desired targets with my recommendation of the strategic adaptive cognition (SAC) approach. Strategies are not just basic planning; fear and discomfort are an essential part of strategy-making and that alone can be an impediment to cognitive leadership which is a critical determinant of leader performance. *Strategies and mechanisms question assumptions.* Questioning strategic assumptions helps to gain commitment from nations. Planning and silo-initiatives do not take that into consideration. It is significant for achieving the goals to concentrate on the key choices to influence the proper decisions and visualize the role of ICT in sustainable development; a vehicle to accelerate and make possible any policy instrument to deter negative effects on accomplishing sustainable development, such as the generation of electronic waste and climate change impacts. On the other hand, ICT has a role, too, as an enabler of efficient resource usage, education, and business operations which is a critical success factor for achieving the SDGs. The agendas of accelerating sustainable development simply using ICTs without an appropriate strategy will prove to be unsustainable. Developing a strategic approach to ICT implementation presents a challenging proposition for the global community, but also creates a universal framework that provides the necessary force to solve the paradigm shift needed to inhibit business agility and synergetic working models; an essential component of growth, development, poverty reduction, environmental stewardship, and sustainability.

Closing the argument, advocacy for sustainable development is emerging strongly. The SDG Index measures 149 countries, comparing their current progress with a baseline measurement taken in 2015. However, more analysis must be devoted to understanding the implications of the existing SDGs programs and to understand the outcomes of the various systems in place and the possible iterations of the initiatives to achieve the SDGs. Most countries lack a concise SDG strategy that encompasses the significant challenges in specific areas such as climate-change mitigation and adaptation, poverty eradication, income inequality, gender equality, education, and a comprehensive economic development strategy. The role of ICT in achieving sustainable development is more prominent and crucial today than ever before. Nevertheless, world leaders play the self-evident role in building the foundations for a strategically imperative and pragmatic approach to solving global issues.

The strategies, frameworks, and mechanisms developed and presented in this book takes the first step in this direction, providing policy- and decision-makers with a deeper understanding of the implications of inaction, the significance of sustainability, pragmatic solutions, forward-thinking leadership, and the intertwined synergies between the economic, social, and environmental pillars that will make us succeed or fail together.

List of Terms

Economic development: The process by which an economy grows or changes and becomes more advanced, especially when both economic and social conditions are improved.

Environmental assets: Entities that provide environmental "functions" or services.

Environmental policy: The commitment of an organization to the laws, regulations, and other policy mechanisms concerning environmental issues.

Inclusive growth: A concept that advances equitable opportunities for economic participants during economic growth with benefits incurred by every section of society. The definition of inclusive growth implies direct links between the macroeconomic and microeconomic determinants of the economy and economic growth.

References

A Framework for Prototyping Telecare Applications – Scientific Figure on Research Gate. Available from: https://www.researchgate.net/The-Services-and-the-messages-of-the-telecare-application_261547675 [Accessed 3 January 2018].

A Progress Review of the Nanotechnology for Solar Energy Collection and Conversion (Solar) NSI. 2015. A Progress Review of the NNI Nanotechnology Signature Initiatives.

Adaptive Cruise Control System Overview. 2005. 5th Meeting of the U.S. Software System Safety Working Group.

Adaptive Signal Control Technologies. 2016. US Department of Transportation.

Allen, Dan. 2014. The Center for Climate and Security. Climate Change and Cyber Threats: Acknowledging the Links.

Allen, W., J. Cruz, and B. Warburton. 2017. How Decision Support Systems Can Benefit from a Theory of Change Approach. *Environmental Management*. https://dx.doi.org/10.1007/s00267-017-0839-y.

Allenby, B. 2006. The Ontologies of Industrial Ecology? *Progress in Industrial Ecology, an International Journal* 3 (1/2): 28–40.

Almeida, Thiago Alexandre Das Neves, Luís Cruz, Eduardo Barata, and Isabel-María García-Sánchez. 2017. Economic Growth and Environmental Impacts: An Analysis Based on a Composite Index of Environmental Damage. *Ecological Indicators* 76: 119–30. doi:10.1016/j.ecolind.2016.12.028

Amorim, H. V., and R. M. Leao. 2005. *Fermentação alcoólica: ciência e Tecnologia*. Piracicaba: Fermentec.

Analysis of Critical Infrastructure, Dependencies and Interdependencies. 2015. Argonne National Laboratory. U.S. Department of Energy Laboratory, Risk and Infrastructure Science Center-Global Security Sciences Division.

Anand, Sudhir, and Amartya Sen. 2000. Human Development and Economic Sustainability. *World Development* 28 (12): 2029–49. doi:10.1016/s0305-750x(00)00071-1.

Anastas, Paul T., and Julie B. Zimmerman. 2003. Design through the 12 Principles of Green Engineering. *Environmental Science & Technology* 37 (5). doi:10.1021/es032373g.

Andreopoulou, Zacharoula S. 2009. Green Informatics: ICT for Green and Sustainability. *Agricultural Informatics* 3(2): 1–8.

Annan, Kofi. 2009. *The Anatomy of a Silent Crisis: Human Impact Report – Climate Change*. Geneva: Global Humanitarian Forum.

Antonella D'Alessandro, Filippo Ubertini, Simon Laflamme, and Annibale L. Materazzi. 2015. Towards Smart Concrete for Smart Cities: Recent Results and Future Application of Strain-Sensing Nanocomposites. *Journal of Smart Cities* 1 (1): 3–14. http://dx.doi.org/10.18063/JSC.2015.01.002.

Appalachian State University. 2015. *The Inverse Square Relationship*. Boone, NC: Appalachian State University Department of Physics and Astronomy.

Arnfalk, P. 2002. Virtual mobility and pollution prevention: The emerging role of ICT based communication in organizations and its impact on travel. Doctoral Dissertation, The International Institute for Industrial Environmental Economics, Lund University, Sweden.

Artioli, Francesca, Michele Acuto, and Jenny Mcarthur. 2017. The Water-Energy-Food Nexus: An Integration Agenda and Implications for Urban Governance. *Political Geography* 61: 215 –23.

Ashby, Simon. 2010. The 2007-09 Financial Crisis: Learning the Risk Management Lessons. University of Nottingham.

Ashford, Nicholas, and Charles Caldart. 2008. *Environmental Law, Policy and Economics. Reclaiming the Environmental Agenda.* Cambridge, MA: MIT Press.

Baker, Judy. 2011. *Climate Change, Disaster Risk and the Urban Poor.* The International Bank for Reconstruction and Development. Washington, DC: The World Bank.

Bakker, Tammo, Ellen Brauers, Gersom Van Der Elst, and Eunice Wangari. 2017. Nanotechnology in Agricultural Production. UN Policy Analysis Branch, Division for Sustainable Development.

Baron, D. and R. Myerson. 1982. Regulating a monopolist with unknown costs. *Econometrica*, 50: 911–30.

Barrionuevo, Juan M., Pascual Berrone, and Joan E. Ricart. 2012. Smart Cities, Sustainable Progress: Opportunities for Urban Development. IESE Business School University of Navarra.

Bartlett, S. et al. 2012. *The State of the World's Children.* New York: United Nations Children's Fund.

Barwise, J.A., L.J. Lancaster, D. Michaels, J.E. Pope, and J.M. Berry. 2011. An Initial Evaluation of a Novel Anesthetic Scavenging Interface. *Anesthesia & Analgesia* 113: 1064–67.

Basden, James, and Michael Cottrell. 2017. How Utilities are Using Blockchain to Modernize the Grid. *Harvard Business Review*.

Bashshur, Rashid L., and Gary W. Shannon. 2009. National Telemedicine Initiatives: Essential to Healthcare Reform. *Telemedicine and e-Health.* doi:10.1089/tmj.2009.9960.

Beck, Matthias, and Sara Melo. 2014. *Quality Management and Managerialism in Healthcare: A Critical Historical Survey.* Basingstoke, UK: Palgrave Macmillan.

Becoming More Sustainable with Standards. 2018. The British Standards Institution (BSI).

Ben-Zeev, Dror. 2016. Mobile Health for All: Public-Private Partnerships Can Create a New Mental Health Landscape. *JMIR Mental Health. Internet Interventions, Technologies and Digital Innovations for Mental Health and Behaviour Change* 3 (2): 2368–7959.

Berkhout, F. and J. Hertin. 2001. *Impacts of Information and Communication Technologies on Environmental Sustainability: Speculations and Evidence.* Brighton, UK: Science and Technology Policy Research.

Bharathi, M.A., and B.P. Vijaya Kumar. 2012. Reverse Game Theory Approach for Aggregator Nodes Selection with Ant Colony Optimization Based Routing in Wireless Sensor Network. *IJCSI International Journal of Computer Science Issues* 9 (6): 292–8.

Bharathidasan, D., and J. Muhibullah. 2006. Solar Cells and Nanotechnology. Dept of Electrical and Electronics Engineering. Greentech College of Engg and Tech for Women. *IOSR Journal of Electrical and Electronics Engineering (IOSR-JEEE)* 8–11.

Bilbao-Osorio, Beñat. 2013. A New Digital Divide Threatens Growth. *World Economic Forum*, 10 April.

Bose, Bimal K. 2017. Artificial Intelligence Techniques in Smart Grid and Renewable Energy Systems—Some Example Applications. *Proceedings of the IEEE* 105 (11): 2262–73. doi:10.1109/jproc.2017.2756596.

Bowles, Samuel, and Sung-Ha Hwang. 2008. Social Preferences and Public Economics: Mechanism Design When Social Preferences Depend on Incentives. *Journal of Public Economics* 92 (8–9): 1811–20. doi:10.1016/j.jpubeco.2008.03.006.

Brugmann, J. 2012. Financing the resilient city. *Environment and Urbanization*, 24 (1): 215–232.

Burney, S. M. Aqil, Nadeem Mahmood, and Zain Abbas. 2010. Information and Communication Technology in Healthcare Management Systems: Prospects for Developing Countries. *International Journal of Computer Applications* 4 (2): 27–32. doi:10.5120/801-1138.

Business Continuity Plan. City of Lincoln, Nebraska. 2009. City of Lincoln Personnel Policy.

Candiello, Antonio, and Agostino Cortesi. 2011. KPI Supported PDCA Model for Innovation Policy Management in Local Government. Dipartimento di Informatica, Universita Ca' Foscari, Venice, Italia.

Caragliu, Andrea, Chiara Del Bo, and Peter Nijkamp. 2009. Smart Cities in Europe. In *3rd Central European Conference in Regional Science – CERS*.

Cash, D.W., W.C. Clark, F. Alcock, N.M. Dickson, N. Eckley, D.H. Guston, J. Jager, and R.B. Mitchell. 2003. Knowledge Systems for Sustainable Development. *Proceedings of the National Academy of Sciences* 100 (14): 8086–91.

Centers for Disease Control and Prevention. 2015. The Four Domains of Chronic Disease Prevention. Atlanta, GA: Centers for Disease Control and Prevention National Center for Chronic Disease Prevention and Health Promotion.

Centre for Research on the Epidemiology of Disasters (CRED) and Institute of Health and Society (IRSS). 2017. *Annual Disaster Statistical Review 2016 – The Numbers and Trends*. Brussels, Belgium: Université catholique de Louvain.

Chancel, L., and T. Piketty. 2015. *Carbon and Inequality: From Kyoto to Paris*. Paris: Paris School of Economics.

Chang, Feng-Cheng, and Huang, Hsiang-Cheh. 2014. A Framework for Prototyping Telecare Applications. *Journal of Information Hiding and Multimedia Signal Processing* 5: 61–71.

Chandrasekhar, C.P. and J. Ghosh. 2001. Information and communication technologies and health in low income countries: The potential and the constraints. *Bulletin of the World Health Organization*, 79 (9): 850–855.

Charron, Nicholas, Lewis Dijkstra, and Victor Lapuente. 2013. Regional Governance Matters: Quality of Government within European Union Member States. *Regional Studies* 48 (1): 68–90. doi:10.1080/00343404.2013.770141.

Chen, Mingxing, Hua Zhang, Weidong Liu, and Wenzhong Zhang. 2014. The Global Pattern of Urbanization and Economic Growth: Evidence from the Last Three Decades. *PLoS One* 9 (8). doi:10.1371/journal.pone.0103799.

Chiasson, M., M. Reddy, B. Kaplan, and E. Davidson. 2007. Expanding Multidisciplinary Approaches to Healthcare Information Technologies: What Does Information Systems Offer Medical Informatics? *International Journal of Medical Informatics* 76 (Suppl 1): S89–97. doi:10.1016/j.ijmedinf.2006.05.010. [PubMed].

Choi, J. and S.Y. Lee. 1999. Factors Affecting the Economics of Polyhydroxyalkanoate (PHAs) Production by Bacterial Fermentation. *Applied Microbiology and Biotechnology* 51: 13–21.

Christensen, Norman L., Ann M. Bartuska, James H. Brown, Stephen Carpenter, Carla D'Antonio, Rober Francis, Jerry F. Franklin, James A. MacMahon, Reed F. Noss, David J. Parsons, Charles H. Peterson, Monica G. Turner, and Robert G. Woodmansee. 1996. The report of the ecological society of America Committee on the Scientific Basis for Ecosystem Management. *Ecological Applications*, 6 (3): 665–691.

Chung, Kim-Sau, and Jeffrey C. Ely. 2002. *Ex-Post Incentive Compatible Mechanism Design*. Department of Economics, Northwestern University.

Ciplet, D., T. Roberts, and M. Khan. 2015. *Power in a Warming World: The New Global Politics of Climate Change and the Remaking of Environmental Inequality*. Cambridge, MA: MIT Press.

Cisco Global Cloud Index: Forecast and Methodology 2016–2021. Cisco and/or Its Affiliates. 2018.

Classic Mechanism Design. 2010. Harvard John A. Paulson School of Engineering and Applied Sciences.

Clinical Decision Support. 2017. *Eight Reasons Why Clinical Decision Support Is More Valuable than Ever*. Elsevier. https://www.elsevier.com/__data/assets/pdf_fi le/0004/509548/himss-cds-brief-2017-final.pdf

Coiera, E., J. Ash, and M. Berg. The unintended consequences of health information technology revisited. *Yearbook of Medical Informatics*, 2016 (1): 163–169. doi:10.15265/IY-2016-014.

Colombo, U. 1988. The Technology Revolution and the Restructuring of the Global Economy. Pp. 23–31 in *Globalization of Technology: International Perspectives*, J.H. Muroyama and H.G. Stever, eds., Washington, DC: National Academy Press.

Commission on Environment and Development. 1987. Our Common Future. United Nations General Assembly as an Annex to document A/42/427 – Development and International Cooperation: Environment. http://www.exteriores.gob.es/ Portal/es/PoliticaExteriorCooperacion/Desarrollosostenible/Documents/In forme%20Brundtland%20(En%20ingl%C3%A9s).pdf.

Committee on Industrialization of Biology: A Roadmap to Accelerate the Advanced Manufacturing of Chemicals; Board on Chemical Sciences and Technology; Board on Life Sciences; Division on Earth and Life Studies; National Research Council. Industrialization of Biology: A Roadmap to Accelerate the Advanced Manufacturing of Chemicals. Washington (DC): National Academies Press (US); 2015 Jun 29. 2, Industrial Biotechnology: Past and Present. https://ww w.ncbi.nlm.nih.gov/books/NBK305455/

Correa, Teresa and Isabel Pavez. 2016. Digital Inclusion in Rural Areas: A Qualitative Exploration of Challenges Faced by People from Isolated Communities. *Journal of Computer-Mediated Communication* 21 (3): 247–63. doi:10.1111/jcc4.12154.

Costa, A. and B. Kallick. 2008. *Learning and Leading with Habits in Mind: 16 Essential Characteristics for Success*. Alexandria, VA: Association for Supervision and Curriculum Development.

Crandall, R., W. Lehr, and R. Litan. 2007. *The Effects of Broadband Deployment on Output and Employment: A Cross-Sectional Analysis of U.S. Data. Issues in Economic Policy 6*. Washington, DC: Brookings Institution.

Cui, Tiansong, Yanzhi Wang, Hadi Goudarzi, Shahin Nazarian, and Massoud Pedram. 2012. Profit Maximization for Utility Companies in an Oligopolistic Energy Market with Dynamic Prices. In *2012 IEEE Online Conference on Green Communications (GreenCom)*, doi:10.1109/greencom.2012.6519621.

Cyber-Security Framework. 2018. Framework for Improving Critical Infrastructure Cyber-Security.; Version 1.1, doi:10.6028/NIST.CSWP.04162018.

Czernich, N., O. Falck, T. Kretschmer, and L. Woessmann. 2011. Broadband infrastructure and economic growth. *The Economic Journal*, 121 (552): 505–532.

de Jong, Carmen. 2009. ICT for the Sustainable Use of Natural Resources with Particular Reference to Water Resources. Environmental Informatics and Industrial Environmental Protection: Concepts, Methods and Tools.

De Schryver A.M., K.W. Brakkee, M.J. Goedkoop, and M.A. Huijbregts. 2009. Characterization Factors for Global Warming in Life Cycle Assessment Based on Damages to Humans and Ecosystems. *Environmental Science & Technology* 43 (6): 1689–95.

Decision Support System (DSS) for Public Policy Making. 2014. European Union under the Seventh Framework Programme. Stockholm University, Department of Computer an (DSV).

Department of Energy and Climate Change. 2012. Statistical Release, 2011 UK Greenhouse Gas Emissions, Provisional Figures and 2010 UK Greenhouse Gas Emissions, Final Figures by Fuel Type and End-user. Available from: http://www.decc.gov.uk/assets/decc/11/stats/climate-change/4817-2011-uk-greenhouse-gas-emissions-provisional-figur.pdf.

Dias de Oliveira, Marcelo E., Burton E. Vaughan, and Edward J. Rykiel Jr. 2005. Ethanol as Fuel: Energy, Carbon Dioxide Balances, and Ecological Footprint. *BioScience* 55 (7): 593–602.

Diesendorf, M. 2000. Sustainability and Sustainable Development. In: Dunphy, D., J. Benveniste, A. Griffiths, and P. Sutton, (eds). *Sustainability: The Corporate Challenge of the 21st Century*, pp. 19–37. Sydney: Allen & Unwin.

Dietz, T., E. Rosa, and R. York. 2012. Environmentally efficient well-being: Is there a Kuznets curve? *Applied Geography*, 32: 21–28.

Digital-Enabled CO_2e Emissions Trajectory towards 2030, Compared to IPCC BAU Scenario. 2016. GeSI, #SystemTransformation report.

Dijkema, G. P. J., and L. Basson. 2009. Complexity and Industrial Ecology: Foundations for a Transformation from Analysis to Action. *Journal of Industrial Ecology* 13 (2): 157–64.

Dixon, B.E., J.M. Hook, and J.J. McGowan. December 2008. Using Telehealth to Improve Quality and Safety: Findings from the AHRQ Portfolio (Prepared by the AHRQ National Resource Center for Health IT under Contract No. 290-04-0016). AHRQ Publication No. 09-0012-EF. Rockville, MD: Agency for Healthcare Research and Quality.

Doarn, Charles, Rifat Latifi, Filip Hostiuc, Raed Arafat, and Claudiu Zoicas. *A Multinational Telemedicine Systems for Disaster Response: Opportunities and Challenges*. Amsterdam, Netherlands: IOS Press, 2017.

Dougherty, Brian, Jules White, and Douglas C. Schmidt. 2012. Model-Driven Auto-Scaling of Green Cloud Computing Infrastructure. *Future Generation Computer Systems* 28 (2): 371–8.

Dukas, R. 1998a. *Cognitive Ecology: The Evolutionary Ecology of Information Processing and Decision Making.* Chicago, IL: Chicago University Press.

Dukas, R. 1998b. Evolutionary ecology of learning. In: Dukas, R. (ed.). *Cognitive Ecology: The Evolutionary Ecology of Information Processing and Decision Making,* pp. 129–174. Chicago, IL: Chicago University Press.

Dunlap, Riley E. and Robert J. Brulle. 2015. *Climate Change and Society: Sociological Perspectives.* New York and Oxford: Oxford University Press.

Eckelman, M.J., Sherman J. 2016. Environmental impacts of the US health care system and effects on public health. *PLoS One* 11 (6): e0157014.

eHealth, Tools and Services. 2006. Needs of the Member States. Report of the WHO Global Observatory for eHealth. World Health Organization.

Electronic Waste. World Health Organization. August 15, 2017. http://www.who.int/ceh/risks/ewaste/en/ [Accessed 30 April 2018].

Elena Raluca Moisescu (Duican), 2015. Analysis of the Relationship between Sustainable Development and Economic Growth. In *International Conference on Marketing and Business Development Journal, The Bucharest University of Economic Studies,* vol. 1 (1), pp. 138–43.

Emergency Radiocommunications. International Telecommunication Union, 2017. https://www.itu.int/en/ITU-R/information/Pages/emergency.aspx

Emerging Issues: The Social Drivers of Sustainable Development. 2014. United Nations Research Institute for Social Development (UNRISD).

Emerging mHealth: Paths for Growth. The Economist Intelligence Unit. PWC. 2015. https://www.pwc.com/gx/en/healthcare/mhealth/assets/pwc-emerging-mhealth-full.pdf.

Environmental Effects and Opportunities Created by ICT. The Impact of ICT on Sustainable Development. 2002. European Information Technology Observatory (E I T O).

Erdmann, L., L. Hilty, J. Goodman, and P. Arnfalk. 2004. The future impact of ICT on environmental sustainability. Technical Report EUR 21384 EN. Institute for Prospective Technological Studies, European Commission Joint Research Center. http://ftp.jrc.es/EURdoc/eur21384en.pdf

European Parliament: Mapping Smart Cities in the EU. 2014. Brussels: European Parliament, Directorate General for Internal Policies.

Everett, Tim, Mallika Ishwaran, Gian Paolo Ansaloni, and Alex Rubin. 2010. Economic Growth and the Environment. Defra- Department for Environment Food and Rural Affairs, Evidence and Analysis Series.

The United Nations Economic Commission for Europe (UNECE). 2017. Exploring the Transformative Potential of Blockchain for Sustainable Development; the 30th UN/CEFACT Forum.

Fedkin, Mark. 2018. Principles of Sustainable Engineering. Technologies for Sustainability Systems, Faculty Department of Mechanical and Nuclear Engineering. PennState.

Fichter, K. 2001. *Sustainable Business Strategies in the Internet Economy. Sustainability in the Information Society.* Zurich: Metropolis Verlag.

Figuères, Caroline M., and Hilde Eugelink. 2014. The Role of ICTs in Poverty Eradication: More Than 15 Years' Experience from the Field. *ICTs and the Millennium Development Goals,* 199–222. doi:10.1007/978-1-4899-7439-6_12.

Figuères, Caroline. 2013. Technical and Social Innovation. International Institute for Communication and Development (IICD). Innovation and Technology for Poverty Eradication.

Figures for the Centers for Disease Control and Prevention. (2010) and the American Diabetes Association, 2009.

Filho, Carlos Massera, Denis F. Wolf, Valdir Grassi Jr, and Fernando S. Osorio. 2014. Longitudinal and Lateral Control for Autonomous Ground Vehicles. *IEEE Intelligent Vehicles Symposium* (IV). doi:978-1-4799-3637-3.

Finlay, Alan, and Adera, Edith. 2012. Application of ICTs for Climate Change Adaptation in the Water Sector: Developing Country Experiences and Emerging Research Priorities. Asian and Pacific Training Centre for Information and Communication Technology for Development (APCICT). https://www.apc.org/sites/default/files/apc_CCW_e%20summary_0.pdf [Accessed 22 April 2018].

Forge, S., Blackman, C., Bohlin, E., Cave, M. September 2009. Green Knowledge Society, an ICT Policy Agenda to 2015 for Europe's Future Knowledge Society. A study for the Ministry of Enterprise, Energy and Communications, Government Offices of Sweden by SCF Associates Ltd. Final Report.

Gagnon, B., and R. Leduc. 2006. Intégration des principes de développement durable dans la conception en ingénierie: la conception durable. *Vecteur Environnement* 39: 31.

Garrity, John, and Robert Pepper. 2015. ICTs, Income Inequality, and Ensuring Inclusive Growth. ICTs, Income Inequality, and Ensuring Inclusive Growth – SSRN Electronic Journal. The Global Information Technology Report. doi:10.2139/ssrn.2588115.

Gartner. 2009. Gartner Estimates ICT Industry Accounts for 2 Percent of Global CO2 Emissions. www.gartner.com/it/page.jsp?id=503867.

Gawande, A.A., M.J. Zinner, D.M. Studdert, and T.A. Brennan. 2003. Analysis of Errors Reported by Surgeons at Three Teaching Hospitals. *Surgery* 133 (6): 614–21. doi:10.1067/msy.2003.169.[PubMed].

Gehrke, I., A. Geiser, A. Somborn-Schulz. 2015. Innovations in Nanotechnology for Water Treatment. *Nanotechnology, Science and Applications* 8: 1–17. doi:10.2147/NSFA.S43773.

Gheorghe, Boaru, and Badita George-Ionut. 2008. Critical Infrastructure Interdependencies, presented at the Conference on Defense Resources Management, Braşov.

Gibson, R.B. 2001. Specification of Sustainability-Based Environmental Assessment Decision Criteria and Implications for Determining 'Significance' in Environmental Assessment. Canadian Environmental Assessment Agency Research and Development Programme, Ottawa.

Gillett, S., W. Lehr, C. Osorio, and S. Marvin. 2006. *Measuring the Economic Impact of Broadband Deployment*. Washington, DC: US Department of Commerce, Economic Development Administration.

GIS Solutions For Environmental Management. 2005. ESRI: Mapping Your Environmental Management Strategy. Redlands, CA: ESRI.

Goldin, C. and L.F. Katz. 1998. The origins of technology-skill complementarity. *Quarterly Journal of Economics*, 113 (3): 693–732.

Goldman, Todd, and Roger Gorham. 2006. Sustainable Urban Transport: Four Innovative Directions. *Technology in Society* 28 (1–2): 261–73. doi:10.1016/j.techsoc.2005.10.007.

Governance for Sustainable Development. 2014. Integrating Governance in the Post-2015 Development Framework. New York: The United Nations Development Programme.

Graf Plessen, Mogens, Daniele Bernardini, Hasan Esen, and Alberto Bemporad. 2018. Spatial-Based Predictive Control and Geometric Corridor Planning for Adaptive Cruise Control Coupled with Obstacle Avoidance. *IEEE Transactions on Control Systems Technology* 26 (1): 38–50. doi:10.1109/tcst.2017.2664722.

Greenberg, A. 2005. ICTs for Poverty Alleviation: Basic Tool and Enabling Sector. Sida, ICT for Development Secretariat, Department for Infrastructure and Economic Cooperation, Article number: SIDA4937en, ISBN 91-586-8429-8.

Grillmayer, Roland. 2002. Landscape Structure Model. In *Environmental Communication in the Information Society – Proceedings Of The 16Th Conference.* doi: ISBN: 3-9500036-7-3.

Groves, T. and J. Ledyard. 1977. Optimal allocation of public goods: A solution to the "Free-Rider" problem. *Econometrica*, 45 (4): 783–809.

Guha-Sapir, D., Ph. Hoyois, P. Wallemacq, and R. Below. 2016. Annual Disaster Statistical Review. The Numbers and Trends. Brussels: CRED.

Gupta, Manoj Kumar. 2013. Applications of ICT in Managing Climate Change. Kathmandu University. Dhulikhel, Kavre, Nepal, April 3. http://dms.nasc.org.np/sites/default/files/documents/ICTforClimate Change.pdf

Hamine, S., E. Gerth-Guyette, D. Faulx, B.B. Green, and A.S. Ginsburg. 2015. Impact of mHealth Chronic Disease Management on Treatment Adherence and Patient Outcomes: A Systematic Review. *Journal of Medical Internet Research* 17 (2): e52. doi:10.2196/jmir.3951.

Hanak, E., and J.R. Lund. 2012. Adapting California's Water Management to Climate Change. *Climate Change* 111: 17–44.

Harvard Business Review. March 2015. Stuff: When Less Is More. *Economics & Society.* hbr.org/2015/03/vision-statement-stuff-when-less-is-more.

Hba, Rachid, and Abdellah El Manouar. 2017. ICT Green Governance: New Generation Model Based on Corporate Social Responsibility and Green IT. *Journal of Data Mining and Digital Humanities.* doi:ISSN 2416-5999.

Health in the green economy. 2011. Co-Benefits to Health of Climate Change Mitigation. Health Care Facilities, Preliminary Findings—Initial Review. http://www.who.int/hia/hgebrief_health.pdf.

Hediger, Werner. 2000. Sustainable Development and Social Welfare. *Ecological Economics* 42 (3): 481–92.

HEI Review Panel on Ultra fine Particles. 2013. Understanding the Health Effects of Ambient Ultrafine Particles. HEI Perspectives. Boston, MA: Health Effects Institute.

Heiskanen, Eva, Minna Halme, Mikko Jalas, Anna Kärnä, and Raimo Lovio. 2001. Dematerialization: The Potential of ICT and Services. Helsinki: The Finnish Environment 533 Ministry of the Environment Environmental Protection Department.

Hellman, H. 1997. *Beyond Your Senses.* New York: Lodestar Books.

Hernan, Galperin, and Maria Fernanda Viecens. 2014. Connected for Development? Theory and Evidence about the Impact of Internet Technologies on Poverty Alleviation. *SSRN Electronic Journal.* doi:10.2139/ssrn.2397394.

Hilty, Lorenz M., Peter Arnfalk, Lorenz Erdmann, James Goodman, Martin Lehmann, and Patrick A. Wäger. 2006. The Relevance of Information and Communication Technologies for Environmental Sustainability – A Prospective Simulation Study. *Environmental Modelling & Software* 21 (11): 1618–629. doi:10.1016/j. envsoft.2006.05.007.

Holler, J., V. Tsiatsis, C. Mulligan, S. Avesand, S. Karnouskos and D. Boyle. 2014. *From Machine-to-Machine to the Internet of Things: Introduction to a New Age of Intelligence.* Waltham : Elsevier.

Holmner, Å., J. Rocklöv, N. Ng, and M. Nilsson. 2012. Climate Change and eHealth: A Promising Strategy for Health Sector Mitigation and Adaptation. *Global Health Action* 5. 10.3402/gha.v5i0.18428.

Hoornweg, Daniel, Lorraine Sugar, Lorena Trejos Gomez, and Claudia Lorena. 2011. Cities and Greenhouse Gas Emissions: Moving Forward. *International Institute for Environment and Development* 23 (1): 207–2013.

Hoornweg, Daniel, Mila, Freire, Marcus J. Lee, Perinaz Bhada-Tata, and Belinda Yuen. 2011. Cities and Climate Change: Responding to an Urgent Agenda. Urban Development Series.World Bank. © World Bank. https://openknowledg e.worldbank.org/handle/10986/2312 License: CC BY 3.0 IGO.

Horrocks, Ivan, and Pratchet, Lawrence. 2009. Democracy and Technology. 'Electronic Democracy: Central Themes and Issues'. Accessed 26 April 2018.

Houghton A. 2011. Health Impact Assessments. A Tool for Designing Climate Change Resilience into Green Building and Planning Projects. *Journal of Green Building* 6: 66–87.

How Digital Solutions will Drive Progress towards the Sustainable Development Goals. Global eSustainability Initiative. SystemTransformation, 2016. http://ges i.org/report/detail/system-transformation.

Human Development Report 2010. 2010. 20th Anniversary Edition. The Real Wealth of Nations: Pathways to Human Development. New York: UNDP.

ICT & SDGs Final Report. 2016. How Information and Communications Technology Can Accelerate Action on the Sustainable Development Goals. The Earth Institute, Columbia University. Accessed 22 January 2018. https://www.eri csson.com/assets/local/news/2016/05/ict-sdg.pdf.

ICT for Data Collection and Monitoring & Evaluation. 2013. Opportunities and Guidance on Mobile Applications for Forest and Agricultural Sectors. International Bank for Reconstruction and Development / International Development Association,World Bank.

ICT Sustainable Development Goals Benchmark. Connecting the Future. Huawei Technologies Co., Ltd, 2017. Accessed February 2018. http://www-file.huaw ei.com/-/media/CORPORATE/PDF/Sustainability/2017-ICT-sustainable-development-goals-benchmark-final-en.pdf.

ICT: Facts and Figures. 2016. ICT Data and Statistics Division Telecommunication Development Bureau International Telecommunication Union, https://www. itu.int/en/ITU-D/Statistics/Documents/facts/ICTFactsFigures2016.pdf.

ICTs as a Catalyst for Sustainable Development. United Nations. Department of Economic and Social Affairs. Sustainable Development Knowledge Platform, 2016.

ICTs' and the Internet's Impact on Job Creation and Economic Growth. The Digital Economy – International Chamber of Commerce, 2012. http://www.iccindiao nline.org/policy_state/28aug/2.pdf.

IEA (International Energy Agency). 2008. Energy Use in Cities. Pp. 179–93 in *World Energy Outlook 2008*, Paris: Organisation for Economic Co-operation and Development/International Energy Agency.

IEA Report on Global Land Transport Infrastructure Requirements. 2013. https://www.iea.org/publications/freepublications/publication/TransportInfrastructureInsights_FINAL_WEB.pdf.

Implementing the Sustainable Development Goals: From Goals to Action. 2017. Asian Development Bank. Publication Stock No. ARM178798-2

Independent Evaluation Group. 2013. World Bank Group Support for Innovation and Entrepreneurship: An Independent Evaluation. World Bank Group, Washington, DC. https://openknowledge.worldbank.org/handle/10986/16665 License: CC BY 3.0 IGO.

Information and communications technologies for inclusive social and economic development. 2014. Commission on Science and Technology for Development. United Nations Economic and Social Council.

Intelligent Transportation Systems. 2018. DSRC: The Future of Safer Driving. Intelligent Transportation Systems Join Program Office. Office of the Assistant Secretary for Research and Technology (OST-R), U.S. Department of Transportation (US DOT).

Intelligent Transportation Systems for Improving Traffic Energy Efficiency and Reducing GHG Emissions from Roadways. 2015. National Center for Sustainable Transportation. Center for Environmental Research and Technology Bourns College of Engineering. Riverside: University of California.

International Data Corporation. 2017. Worldwide Semiannual Public Cloud Services Spending Guide.

International Telecommunication Union (ITU) and the United Nations Educational, Scientific and Cultural Organization (UNESCO). 2012. The Broadband Bridge: Linking ICT with Climate Action for a Low-Carbon Economy. The Broadband Commission for Digital Development. http://www.broadbandcommission.org/Documents/publications/BD-bbcomm-climate.pdf [Accessed 16 July 2017].

Investment for Sustainable Development. 2015. OECD and Post-2015 Reflections. The Organisation for Economic Co-operation and Development.

IPCC. 2014. *Climate Change 2014: Impacts, Adaptation, and Vulnerability. Contribution of Working Group II to the Fifth Assessment Report of the Intergovernmental Panel on Climate Change*. Cambridge, UK: Cambridge University Press.

Isenmann, Ralf, and Konstantin Chernykh. 2009. The Role of ICT in Industrial Symbiosis Projects – Environmental ICT Applications for Eco-industrial Development. Environmental Informatics and Industrial Environmental Protection: Concepts, Methods and Tools. Fraunhofer Institute for Systems and Innovation Research (ISI), doi:10.18411/a-2017- Shaker Verlag. ISBN: 978-3-8322-8397-1023.

ISO 37120 Standard on City Indicators. 2014. How they help city leaders set tangible targets, including service quality and quality of life.

ISO 22301 Societal Security. 2017. Societal security: Business Continuity Management Systems Requirements. Australasian Council of Security Professionals.

ISO37101 Sustainable Development in Communities. 2016. International Organization for Standardization, doi:ISBN 978-92-67-10680-9.

ITU. 2009. ICTs and Climate Change, background paper for the ITU Symposium on ICTs and Climate Change, Quito, Ecuador, 8–10 July.

Jawahir, L.S., and Ryan Bradley. 2016. Technological Elements of Circular Economy and the Principles of 6R-Based Closed-loop Material Flow in Sustainable Manufacturing. In *Procedia CIRP* 40: 103–8. doi:10.1016/j.procir.2016.01.067.

Jensen, M.E. 2007. *Evidence and Implications of Recent Climate Change.* Netherlands: Springer.

John P. Kotter. 1996. *Leading Change: An Action Plan from the World's Foremost Expert on Business Leadership.* Harvard Business Press, ISBN: 0875847471.

Jorgenson, A. 2014. Economic Development and the Carbon Intensity of Human Well-being. Nature Climate Change 4: 186–89.

Jorgenson, Andrew, Juliet Schor, and, Xiaorui. Huang. 2017. Income Inequality and Carbon Emissions in the United States: A State-Level Analysis, 1997–2012. *Ecological Economics* 134: 40–48. 10.1016/j.ecolecon.2016.12.016.

Kant, Lalit, and Sampath K Krishnan. 2010. Information and Communication Technology in Disease Surveillance, India: A Case Study. *BMC Public Health* 10 (Suppl 1): S11. PMC. Web. 5 May 2018.

Kemp, R. and P. Martens. 2007. Sustainable development: How to manage something that is subjective and never can be achieved? *Sustainability: Science, Practice and Policy,* 3 (2): 5–14. doi:10.1080/15487733.2007.11907997.

Kemp, R., S. Parto and R. Gibson. 2005. Governance for sustainable development: Moving from theory to practice. *International Journal of Sustainable Development,* 8 (1–2): 13–30.

Kendall, E., H. Muenchberger, C. Ehrlich, and K. Armstrong. 2009. *Supporting Self-Management in General Practice: An Overview.* Queensland, Australia: Griffith University and General Practice.

Kenneth S. Johnson 2017. Solid-State Sensors, Actuators and Microsystems. *IEEE.* doi:10.1109/TRANSDUCERS.2017.7993975

Khalil, Mohsen, and Charles Kenny. 2008. *The Next Decade of ICT Development: Access, Applications and the Forces of Convergence.* Washington, DC: The World BankGroup.

Khanka, S.S. 2015. *Business Ethics and Corporate Social Responsibility.* New Delhi: S. Chand Publishing.

Koutroumpis, P. 2009. The economic impact of broadband on growth: A simultaneous approach. *Telecommunications Policy,* 33 (9): 471–485.

Knight, K. 2013. Interview by Marolla, C. [Personal Interview]. Risk Management ISO 31000: Implementing ISO 31000 for Major Cities to Combat Climate Change Impacts on the urban Poor's Health.

Kotter, John P., and Leonard A. Schlesinger. 2013. Choosing Strategies for Change. *Harvard Business Review,* July–August.

Kreps, G.L. 2017. The relevance of health literacy to mHealth. *Studies in Health Technology and Informatics,* 240: 347–55.

Kristof, Taryn, Mike Lowry, and G. Scott Rutherford. 2005. Assessing the Benefits of Traveler and Transportation Information Systems. Washington State Transportation Commission.

Kriza, Máté. 2017. Circular Economy in Support of Resource Efficiency, Climate Action and Sustainable Development. In *25th OSCE Economic and Environmental Forum "Greening the Economy and Building Partnerships for Security in the OSCE Region".*

Kuzma J., and P. VerHage. 2006. *Nanotechnology in Agriculture and Food Production: Anticipated Applications.* Washington, DC: Woodrow Wilson International.

Kwak, Jaeyoung, Byungkyu Park, and Jaesup Lee. 2012. Evaluating the Impacts of Urban Corridor Traffic Signal Optimization on Vehicle Emissions and Fuel Consumption. *Transportation Planning and Technology* 35 (2): 145–60. doi:10.1080/03081060.2011.651877.

Kyriacou, E., S. Pavlopoulos, A. Berler, M. Neophytou, A. Bourka, A. Georgoulas, A. Anagnostaki, D. Karayiannis, C. Schizas, C. Pattichis, A. Andreou, and D. Koutsouris. 2003. *Biomedical Engineering* Online. 2: 7.[PubMed].

Lamberti, Fabrizio, Bartolomeo Montrucchio, Andrea Sanna, and Claudio Zunino. 2008. A Web-based Architecture Enabling Multichannel Telemedicine Applications. Systemics, Cybernetics and Informatics. Dipartimento di Automatica e Informatica, Politecnico di Torino, corso Duca degli Abruzzi 24, I-10129 Torino (Italy). Vol. 1 (1).

Leape, L.L., and D.M. Berwick. 2005. Five years after to Err Is Human: What Have We Learned? *JAMA* 293 (19): 2384–90. doi:10.1001/jama.293.19.2384.[PubMed]

Leaving No One Behind. 2016. The Sustainable Development Goals Report, 48–49. doi:10.18356/d2c06f02-en.

Ledyard, J.O. 1991. Market failure. In: Eatwell, J., M. Milgate, and P. Newman (eds). *The World of Economics*, pp. 407–412. London: Palgrave Macmillan.

Levy, J.I., J.J. Buonocore, and K. von Stackelberg. 2010. Evaluation of the Public Health Impacts of Traffic Congestion: A Health Risk Assessment. *Environmental Health* 9: 65. doi:10.1186/1476-069X-9-65.

Loerincik, Y. 2006. Environmental Impacts and Benefits of Information and Communication Technology Infrastructure and Services, Using Process and Input-Output Life Cycle Assessment. Thesis, École Polytechnique Fédérale de Lausanne.

Logan, D. and H. Fischer-Wright. 2009. Micro strategies: The key to successful planning in uncertain times. *Leader to Leader*, 54: 45–52.

Lowry, G.V., E.M. Hotze, E.S. Bernhardt, D.D. Dionysiou, J.A. Pedersen, M.R. Wiesner, and B. Xing. 2010. Environmental Occurrences, Behavior, Fate, and Ecological Effects of Nanomaterials: An Introduction to the Special Series. *Journal of Environmental Quality* 39: 1867.

MacLean, D. 2008. ICTs, adaptation to climate change and sustainable development at the margins, ITU Climate Change Symposium London, 17–18 June.

Madden, Peter, and Ilka Weißbrod. 2008. Connected ICT and Sustainable Development. Forum for the Future – Action for a Sustainable World, Overseas House, 19–23 Ironmonger Row, London, EC1V 3QN.

Maezawa, Y., Y. Hatakeyama, M. Saito and F. Hirota. 2018. Conservation of Biodiversity by Making Use of ICT. *Fujitsu.com*. https://www.fujitsu.com/global/documents/about/resources/publications/fstj/archives/vol50-4/paper07.pdf [Accessed 26 April 2018].

Management of Continual Improvement for Facilities and Activities: A Structured Approach. 2006. International Atomic Energy Agency. Nuclear Power Engineering Section International Atomic Energy Agency. doi:ISBN 92–0–102906–3.

Manojlovich, M., J. Adler-Milstein, M. Harrod, A. Sales, T. P. Hofer, S. Saint, and S. L. Krein. 2015. The Effect of Health Information Technology on Health Care Provider Communication: A Mixed-Method Protocol. *JMIR Research Protocols* 4 (2): e72. http://doi.org/10.2196/resprot.4463.

Markovic, Dragan S., Dejan Zivkovic, Dragan Cvetkovic, and Ranko Popovic. 2012. Impact of Nanotechnology Advances in ICT on Sustainability and Energy Efficiency. *Renewable and Sustainable Energy Reviews* 16 (5): 2966–72. doi:10.1016/j. rser.2012.02.018.

Marolla, Cesar. 2017. *Climate Health Risks in Megacities: Sustainable Management and Strategic Planning.* Boca Raton, FL: CRC Press/Taylor & Francis Group.

Maskin, E. A. 2014. Dialogue with Nobel Laurate Eric Maskin: Successful Negotiating and Financial Techniques for Dealmaking, Investment & Law. Directors Roundtable, October 9, 2014, New York.

Maskin, Eric. Notes on Auctions for Pollution Reduction. Personal communication, March 5, 2018).

Masood, Bilal, Sobia Baig, and Abd-ur-Rehman Raza. 2015. Role of ICT in Smart Grid. *Journal of Electrical Engineering*, Available from: http://www.jee.ro/covers/art.p hp?issue=WO1352871945W50a33009dea24 [Accessed 8 April 2018]

Mayo, J.W. and S. Wallsten. 2011. From network externalities to broadband growth externalities: A bridge not yet built. *Review of Industrial Organization*, 38 (2): 173–190.

Measuring the Relationship between ICT and the Environment. 2009. *Organisation for Economic Co-operation and Development.* [online] Available at: http://www.oecd. org/internet/ieconomy/43539507.pdf [Accessed 20 June 2017].

Mechanism Design: Some Definitions and Results. 2016. London: Department of Economics, UCL (University College London).

Meera, Shaik N., Anita Jhamtani, and D.U.M. Rao. 2004. Information and Communication Technology in Agricultural Development: A Comparative Analysis of Three Projects from India. Agricultural Research & Extension Network. The Agricultural Research and Extension Network Is Sponsored by the UK Department for International Development (DFID). doi:10.1107/s010827 0113015370/sk3488sup1.cif.

Menashri, Harel, and Gil Baram. 2015. Critical Infrastructures and Their Interdependence in a Cyber Attack – The Case of the U.S. *Military and Strategic Affairs* 7: (1): 79–100.

Mensah, Peter, Yuri Merkuryev, and Longo Francesco. 2015. Using ICT in developing a resilient supply chain strategy. *Procedia Computer Science*, 43: 101–108.

McKinsey Global Institute. 2012. *Manufacturing the Future: The Next Era of Global Growth and Innovation.* n.p.: McKinsey Global Institute.

McLeod, K.L., J. Lubchenco, S.R. Palumbi, and A.A. Rosenberg. 2005. Scientific Consensus Statement on Marine Ecosystem-Based Management. http://mar ineplanning.org/wp-content/uploads/2015/07/Consensusstatement.pdf.

McMichael, A.J., D.H. Campbell-Lendrum, C.F. Corvalán, K.L. Ebi, A. Githelo, J.D. Scheraga, and A. Woodward. 2003. *Climate Change and Human Health – Risks and Responses.* Geneva: World Health Organization.

mHealth New Horizons for Health through Mobile Technologies. 2011. Based on the Findings of the Second Global Survey on EHealth. Global Observatory for EHealth Series – World Health Organization3. ISBN 978 92 4 156425 0.

Millennium Development Goals (MDGs). 2015. *United Nations.* http://www.un.org/ millenniumgoals/ [Accessed 10 July 2017].

Milliman Research Report. 2011. Milliman Medical Index: Healthcare Costs for American Families Double in Less Than Nine Years. http://www.ucs-edu.net/ cms/wp-content/uploads/2014/03/milliman-medical-index-2011.pdf.

Minges, Michael. 2015. Exploring the Relationship between Broadband and Economic Growth. World Development Report, Digital Dividends. World Bank. http://pubdocs.worldbank.org/en/391452529895999/WDR16-BP-Exploring-the-Relationship-between-Broadband-and-Economic-Growth-Minges.pdf.

MIT Center for Transportation and Logistics. 2017. Mapping Supply Chain Response to Cyber Attacks. Massachusetts Institute of Technology.

Mitchell, Shane, and Wolfgang Wagener. 2010. *The Connected Bus: Connected and Sustainable Mobility Pilot*. San Francisco, CA: Cisco Internet Business Solutions Group.

Mobile Coaching and Online Decision Support for Diabetes. Ambient Assisted Living. ARUP, 2014.

Monyooe, Lebusa, and Steve Ledwaba. 2004. *Information and Communication Technologies for Women Empowerment: South Africa's Unfinished Agenda*. National Research Foundation, Institutional Repository. http://www.itdl.org/journal/sep_04/article01.htm.

Mukhopadhyay S.S. 2014. Nanotechnology in Agriculture: Prospects and Constraints. *Nanotechnology, Science and Applications* 7: 63–71. doi:10.2147/NSA.S39409.

Mulligan, Catherine. 2014. ICT & the Future of Transport. Imperial College London. Industry Transformation – Horizon Scan: ICT & the Future of Transport.

Myerson, Roger B. 1981. Optimal Auction Design, *Mathematics Operations Research* 6: 58–73.

Nabyonga-Orem, Juliet. 2017. Monitoring Sustainable Development Goal 3: How Ready Are the Health Information Systems in Low-Income and Middle-Income Countries? *BMJ Global Health* 2 (4): e000433. doi:10.1136/bmjgh-2017-000433.

Nanoforum Nanotechnology in Agriculture and Food. 2006. A Nanoforum report. Available from: http://urlm.co/www.nanoforum.org.

Narahari, Y. 2009. *Game Theoretic Problems in Network Economics and Mechanism Design Solutions*. London: Springer.

National Institute of Standards and Technology. 2018. Framework for Improving Critical Infrastructure Cyber-Security. Gaithersburg, MD: NIST.

National Institute of Standards and Technology. 2018. Cyber-Security Framework Version 1.1. Gaithersburg, MD: NIST, Department of Commerce.

National Ocean Service. 2015. How Are Satellites Used to Observe the Ocean? National Ocean Service (NOAA), US Department of Commerce. National Oceanic and Atmospheric Administration, Department of Commerce. https://oceanservice.noaa.gov/facts/satellites-ocean.html [Accessed 3 January 2018].

Nelson, Richard R. 1999. *Harnessing Science and Technology for America's Economic Future: National and Regional Priorities*. Washington, DC: National Academy Press.

Nicogossian A.E., and C.R. Doarn. 2011. Armenia 1988 Earthquake and Telemedicine: Lessons Learned and Forgotten. *Telemedicine Journal and e-Health* 17: 741–5. [PubMed].

OECD. 2015. The OECD Corporate Governance Principles. The International Corporate Governance System. doi:10.1057/9781137360014.0010.

Omer A.M. 2008. Renewable Building Energy Systems and Passive Human Comfort Solutions. *Renewable and Sustainable Energy Reviews* 12: 1562–87.

Omoruyi, Eke and Onafalujo Kunle. 2011. Effects of climate change on health risks in Nigeria. *Asian Journal of Business and Management Sciences*, 1 (1): 204–215.

Ono, Takayuki, Kenichi Iida, and Seiya Yamazaki. 2017. Achieving Sustainable Development Goals (SDGs) Through ICT Services. *Fujitsu Scientific & Technical Journal* 53 (6): 17–22.

Optimal Signal Control with Multiple Objectives in Traffic Mobility and Impacts. 2009. KTH Royal Institute of Technology.

Ortman, Jennifer M.Victoria A. Velkoff, and Howard Hogan. 2014. An Aging Nation: The Older Population in the United States. Population Estimates and Projections. Current Population Reports. U.S. Department of Commerce Economics and Statistics Administration U.S. CENSUS BUREAU.

Ospina, Angelica Valeria, and Richard Heeks. 2010. Unveiling the Links between ICTs & Climate Change in Developing Countries: Centre for Development Informatics - University of Manchester. https://idl-bnc-idrc.dspacedirect.org/bitstream/handle/10625/44414/130854_.pdf?sequence=1.

Oxford Dictionary. 2018. Definitions: Informatics. https://en.oxforddictionaries.com/definition/informatics.

Parisi, C., M. Vigani, and E. Rodríguez-Cerezo. 2014. Proceedings of a workshop on "Nanotechnology for the agricultural sector: from research to the field." Publications Office of the European Union, Luxembourg.

Parry, Martin, Osvaldo Canziani, Jean Palutikof, Paul van der Linden, and Clair Hanson. 2008. *Climate Change 2007: Impacts, Adaptation and Vulnerability. Contribution of Working Group II to the Fourth Assessment Report of the Intergovernmental Panel on Climate Change Published for the Intergovernmental Panel on Climate Change*. New York: Cambridge University Press.

Pawlak, Jacek, Scott Le Vine, John Polak, Aruna Sivakumar, and Johanna Kopp. 2015. ICT and PHYSICAL MOBILITY. State of Knowledge and Future Outlook. Institute for Mobility Research. https://www.ifmo.de/files/publications_content/2015/ifmo_2015_ICT and Physical Mobility_en.pdf.

Pepper, Robert. 2015. The Digital Tipping Point and the Paradox of Income Inequality.

Pepper, Robert, and John Garrity. ICTs, Income Inequality, and Ensuring Inclusive Growth. *CISCO*, 2015. Accessed April 2018. https://alln-extcloud-storage.cisco.com/ciscoblogs/GITR-2015-Cisco-Chapter.pdf.

Pigliucci, M. 2001. *Phenotypic Plasticity: Beyond Nature and Nurture*. Baltimore, MD: Johns Hopkins Press.

Pojasek, Robert. 2010. Sustainability: The Three Responsibilities. *Environmental Quality Management* 19 (3): 87–94

Policy Coherence for Sustainable Development in the SDG Framework. Shaping Targets and Monitoring Progress. 2015. The Organisation for Economic Co-operation and Development.

Population Ageing and Sustainable Development. 2014. United Nations Department of Economics and Social Affairs, Population Division.

Porter, M. and M. Kramer. 2011. Creating shared value. *Harvard Business Review*, 89 (1–2): 62–77.

Proposed ICT Indicators for the SDG Monitoring Framework. 2015. International Telecommunications Union (ITU), https://www.itu.int/en/ITU-D/Statistics/Documents/intlcoop/sdgs/ITU-ICT-indicators-for-the-SDGs-Sept2015.pdf.

Qalyoubi, Rula. 2012. *The Seven Steps towards Green Governance*. Network for Green Economies Indicator. Ecologic Institute.

Qiang, C.Z. and C.M. Rossotto. 2009. Economic impacts of broadband. In: World Bank (ed.). *Information and Communications for Development 2009: Extending Reach and Increasing Impact*, pp. 35–50. Washington, DC: World Bank.

Quest Sustainability Assessor. 2017. Sustainability Benchmarking and Measurement.

Quinn, C. C., Swasey, K. K., Crabbe, J. C. F., Shardell, M. D., Terrin, M. L., Barr, E. A., and Gruber-Baldini, A. L. 2017. The Impact of a Mobile Diabetes Health Intervention on Diabetes Distress and Depression Among Adults: Secondary Analysis of a Cluster Randomized Controlled Trial. *JMIR mHealth and uHealth* 5 (12): e183. http://doi.org/10.2196/mhealth.8910.

Radermacher, F., 1999, Telework: Its Role in Achieving a Sustainable Global Economy. www.faw.uni-ulm.de/deutsch/publikationen/radermacher/telework.html.

Ramesh, G., Vishnuprasad Nagadevara, Gopal Naik, Anil Suraj, and Indian Institute of Management, Bangalore. 2014. *Public Private Partnerships*. New York: Routledge, Taylor & Francis Group.

Raupach M.R., G. Marland, P. Ciais, et al. 2007. Global and Regional Drivers of Accelerating CO_2 Emissions. *Proceedings of the National Academy of Sciences of the USA* 104 (24): 10288–93.

Richardson, D. 1996. *The Internet and Rural Development: Recommendations for Strategy and Activity – Final Report*. Rome: Sustainable Development Department of the Food and Agriculture Organization of the United Nations.

Rileyand, John G., and William F. Samuelson. 1981. Optimal Auctions, *American Economic Review* 71: 381–92.

Roberts, J.T. and B.A. Parks. 2017. *Climate of Injustice: Global Inequality, North-South Politics, and Climate Policy*. Cambridge, MA: MIT Press.

Robèrt, K.-H., B. Schmidt-Bleek, J. Aloisi De Larderel, G. Basile, J.l. Jansen, R. Kuehr, P. Price Thomas, M. Suzuki, P. Hawken, and M. Wackernagel. 2002. Strategic Sustainable Development—Selection, Design and Synergies of Applied Tools. *Journal of Cleaner Production* 10 (3): 197–214. doi:10.1016/s0959-6526(01)00061-0.

Roco, M.C. 1999. Towards a US National Nanotechnology Initiative. *Journal of Nanoparticle Research* 1: 435–38.

Roebuck, M.C., et al. 2011. Medication Adherence Leads to Lower Health Care Use and Costs Despite Increased Drug Spending, *Health Affairs (Millwood)* 30 (1): 91–9.

Rollason, V. 2010. Applying the ISO 31000 risk assessment framework to coastal zone management. *NSW Government's Sea Level Rise Policy Statement, Coastal Planning Guideline: Adapting to Sea Level Rise*, 4–13.

Rollason, V. et al. 2011. Applying the ISO 31000 risk assessment framework to coastal zone management. *Australian Standard for Risk*, 4–5.

Romeiro, Ademar Ribeiro. 2012. Sustainable Development: An Ecological Economics Perspective. *Estudos Avançados* 26 (74): 65–92.

Roeth H., and Wockek L. 2011. *ICT and Climate Change Mitigation in Emerging Economies.*, Manchester, UK: Centre for Development Informatics, Institute for Development Policy and Management, SED.

Roulstone, Alan. 2016. *Disability and Technology: An Interdisciplinary and International Approach*. London: Palgrave Macmillan.

Rutkąuskiene, D., and D. Gudoniene. 2015. *Methods and Technologies for ICT Workers Virtual Mobility*. In V. Uskov R., Howlett L., Jain (eds), *Smart Education and Smart e-Learning. Smart Innovation, Systems and Technologies*, Vol. 41. Cham: Springer.

Ryan S.M., and C.J. Nielsen. 2010. Global Warming Potential of Inhaled Anesthetics: Application to Clinical Use. *Anesthesia & Analgesia* 111: 92–98.

Sacks D.B. 2012. Measurement of Hemoglobin A1c: A New Twist on the Path to Harmony. *Diabetes Care* 35 (12): 2674–2680. doi:10.2337/dc12-1348.

Sakson, Steve and George Whitmore. 2013. Communications Strategy: A Vital (but Often Overlooked) Element in Lean-Management Transformations. *McKinsey Operations Extranet.* https://operations-extranet.mckinsey.com/article/communications-strategy.

Samii, Roxanna. 2008. Role of ICTs as Enablers for Agriculture and Small-Scale Farmers. International Fund for Agricultural Development (IFAD).

Saving Carbon, Improving Health, NHS carbon reduction strategy for England. 2009. Available from:http://www.sdu.nhs.uk/documents/publications/1237308334_qylG_saving_carbon,_improving_health_nhs_carbon_reducti.pdf.

Schenberg, A. 2010. Biotechnology and sustainable development. *Estudos Avançados,* 24 (70): 7–17.

Schenberg, Guerrini, and Ana Clara. 2010. Biotechnology and Sustainable Development. Estudos Avançados. NAP-Biotechnology, Institute of Biomedical Sciences, USP. 24 (70).

Schiller, Preston L., and Jeffrey R. Kenworthy. 2018. *An Introduction to Sustainable Transportation: Policy, Planning and Implementation.* Abingdon, Oxon: Routledge.

Science and Technology for Sustainable Development. 2003. Commission on Science and Technology for Sustainable Development in the South. COMSATS.

Scott R.E., T. Perverseff, N. Lefebre. Reducing Environmental Impact: An Example of How e-Health Can Reduce Environmental Impact and Concomitantly Improve Health. In *Proceedings of The 1st Annual Conference on e-Health: The Virtual Dimensions of Health and Environment—Empower, Enhance, Enforce,* 2009. April, pp. 101–10.

Sen, Arunava. 2007. The theory of mechanism design: An overview. *Economic and Political Weekly,* 42 (49): 8–13. www.jstor.org/stable/40277014.

Serena, D. Pedro. 2015. *Working Guide on "Nanotechnology for Urban Revolution: Smart Cities".* Fundacion San Patricio.

Shafiq, Farhan, and Ahsan, Kamran. 2014. An ICT Based Early Warning System for Flood Disasters in Pakistan. *Research Journal of Recent Sciences* 3: 108–18.

Shimizu, K. and M. Hitt. 2004. Strategic flexibility: Organizational preparedness to review ineffective strategic decisions. *Academy of Management Executive,* 18 (4): 44–59.

Short, C.A., K.J. Lomas, A. Woods. 2004. Design strategy for Low-Energy Ventilation and Cooling within an Urban Heat Island. *Building Research & Information* 32: 187–206.

Significant Milestone for Smart City Development – Standards Organizations Agree to Work Together to Move Cities to Greater Smartness. 2017. International Organization for Standardization.

SMART 2020: Enabling the Low Carbon Economy in the Information Age. The Climate Group on behalf of the Global eSustainability Initiative (GeSI), 2008. http://gesi.org/files/Reports/Smart%202020%20report%20in%20English.pdf, [Accessed 2017].

Smart Cities and Infrastructure. 2016. Commission on Science and Technology for Development. United Nations Economic and Social Council.

Smart Cities Preliminary Report. International Organization for Standardization, 2014. ISO/IEC JTC 1, Information Technology.

Smart Cities: How Rapid Advances in Technology Are Reshaping Our Economy and Society. 2015. Deloitte.

Smart Leadership for Smart Cities. Leadership Foundations for Smart City Practitioners, 2016. Birmingham Business School.

Smart Sensor Networks: Technologies and Applications for Green Growth. 2009. Organization for Economic Co-Operation and Development (OECD). https://www.oecd.org/sti/ieconomy/44379113.pdf.

SMARTer2030.gesi.org. (2015). SMARTer 2030. ICT Solutions for 21st Century Challenges.

Smil, Vaclav. 2013. *Making the Modern World: Materials and Dematerialization*. Hoboken, NJ: John Wiley and Sons.

Societal Security Emergency Management. Requirements for Incident Response. 2011. International Organization for Standardization.

Spinak, Abby, and Federico Casalegno. Sustainable and Equitable Urbanism. The Role of ICT in Ecological Culture Change and Poverty Alleviation. IGI Global, 2012. doi:10.4018/978.

Sprague, R.H., Jr., and E.D. Carlson. 1982. *Building Effective Decision Support Systems*. Englewood, Cliffs, NJ: Printice-Hall, 329 p.

Special Issue on Sustainability and Energy. 2017. *Science*, 315: 721–896.

Srivatsan, T. S., and T. S. Sudarshan . 2016. *Additive Manufacturing: Innovations, Advances, and Applications*. Boca Raton: CRC Press/Taylor & Francis Group.

Stantcheva, S. 2017. Introduction to Graduate Public Economics. Lecture, Harvard University, Massachusetts.

Steventon A., M. Bardsley, H. Doll, E. Tuckey, and S.P. Newman. 2014. *BMC Health Services Research*. 14: 334. doi:10.1186/1472-6963-14-334. PMID: 25100190.

Stowe, S., and S. Harding. 2010. Telecare, Telehealth and Telemedicine. *European Geriatric Medicine* 1 (3): 193–7.

Strategic ICT. 2012. How Is ICT Governance Used in Higher Education? The University of Nottingham. I HATe U

Sustainability and Artificial Intelligence Lab. 2018. Stanford University.

Street, A., R. Sustich, J. Duncan, and N. Savage. 2014. *Nanotechnology Applications for Clean Water: Solutions for Improving Water Quality*, 2nd Ed. Oxford: Elsevier.

Suharno, S. 2010. Cognitivism and its implications on the second language learning. *Parole: Journal of Linguistics and Education* 1: 72–96.

Sustainable Consumption and Production Indicators for the Future SDGs UNEP Discussion Paper – March 2015. 2015. United Nations Environment Programme. https://www.iisd.org/sites/default/files/publications/sustainable-consumption-production-indicators-future-sdgs_0.pdf.

Sustainable Consumption Production. United Nations Sustainable Development Goal 12. 2016.

Sustainable Development Challenges. World Economic and Social Survey. 2013. Department of Economic and Social Affairs, United Nations.

Sustainable Development Goal 8. 2017. Promote Sustained, Inclusive and Sustainable Economic Growth, Full and Productive Employment and Decent Work for All. UNSTATS, United Nations.

Sustainable Development Goals ICT Playbook. From Innovation to Impact. 2015. Intel Corporation.

Sustainable Development Goals. United Nations. http://www.un.org/sustainabledevelopment/sustainable-development-goals/ [Accessed 3 February 2018].

Sustainable Development Knowledge Platform. 2015. Partnerships Key to Implementing New Sustainable Development Agenda, General Assembly Summit. United Nations Department of Economic and Social Affairs.

Sustainable Science Institute. 2014. ICT for Health Project. Bill & Melinda Gates Foundation and Fundación Carlos Slim para la Salud.

Sutcliffe, K.M., E. Lewton, M.M. Rosenthal. 2004. Communication Failures: An Insidious Contributor to Medical Mishaps. *Academic Medicine* 79 (2): 186–94.[PubMed].

Tan, Joseph. 2012. *E-Health Care Information Systems: An Introduction for Students and Profess.* San Francisco, CA: Jossey-Bass.

Tang R., Y. Shi, Z. Hou, and L. Wei. 2017. Carbon Nanotube-Based Chemiresistive Sensors. P. 882 in T.M. Swager, K. Mirica, eds. *Sensors* (Basel, Switzerland). Vol. 17 (4). doi:10.3390/s17040882.

Tang, L., R, Ii, K. Tokimatsu, N. Itsubo. 2015 Development of Human Health Damage Factors Related to CO_2 Emissions by Considering Future Socioeconomic Scenarios. *International Journal of Life Cycle Assessment.* 1–12.

Telehealth to Digital Medicine: How 21st Century Technology Can Benefit Patients. 2014. House Energy and Commerce Committee. Subcommittee on Health.

Testa MA, et al. 1998. Health Economic Benefits and Quality of Life During Improved Glycaemic Control in Patients with Type 2 Diabetes Mellitus. *JAMA* 280 (17): 1490–96.

Thatcher, Andrew, and Mbongi Ndabeni. 2011. A Psychological Model to Understand E-Adoption in the Context of the Digital Divide. In: Jacques Steyn and Graeme Johanson, eds. *ICTs and Sustainable Solutions for the Digital Divide: Theory and Perspectives,* 127–50. Hershey, NY: Information Science Reference.

The British Standards Institution. 2015. *Smart Cities Overview – Guide.* London: BSI Department of Business and Innovation.

The Fourth Industrial Revolution for the Earth. Harnessing Artificial Intelligence for the Earth. 2018. PricewaterhouseCoopers. https://www.pwc.com/gx/en/sustainability/assets/ai-for-the-earth-jan-2018.pdf.

The Impact of Smartphone Applications on the Mobile Health Industry (vol. 2), Mobile Health Market, Report 2011–2016, research2guidance, 9 January 2012.

The Innovation Policy Reform. 2013. World Bank Group and the Organisation for Economic Co-operation and Development (OECD).

The United Nations Development Programme (UNDP). 2018. The Future is Decentralised. Block Chains, Distributed Ledgers, & The Future of Sustainable Development. http://www.undp.org/content/dam/undp/library/innovation/The-Future-is-Decentralised.pdf.

The Transforming Mobility Ecosystem: Enabling in Energy-Efficient Future. 2017. Office of Energy Efficiency and Renewable Energy. US Department of Energy, doi:10.2172/1416168.

The World Bank. 2018. What Are Public Private Partnerships? Public–Private-Partnership in Infrastructure Resource Center.

The World Bank; World Bank Institute Economic Policy and Poverty Reduction Division.

Thöni, Andreas, and A Min Tjoa. 2015. Information Technology For Sustainable Supply Chain Management: A Literature Survey. *Enterprise Information Systems* 11 (6): 828–58. doi:10.1080/17517575.2015.1091950.

TL 9000 – Quality Certifications. 2010. CISCO.

TL 9000 Quality Management System. The Telecom Management Quality System, 2017.

Tongia, Rahul, Eswaran Subrahmanian, and V. S. Arunachalam. 2005. *ICT for Sustainable Development. Defining a Global Research Agenda*. Bangalore: Institute for Software Research International School of Computer Science Carnegie Mellon University.

Toolkit on Environmental Sustainability for the ICT Sector. International Telecommunication Union, 2012. https://www.itu.int/dms_pub/itu-t/oth/4B/01/T4B010000060002PDFE.pdf [Accessed 28 April 2018].

Translating Technologies into Competitive, Validated & Manufacturable Products to Impact Quality of Life. 2016. Report on the 10th Concertation & Consultation Workshop on Micro-Nano-Bio- Systems-MNBS 2016.

Trevor, Jonathan, and Barry Varcoe. 2017. How Aligned Is your Organization? *Harvard Business Review*, February 7.

Troll, Carl. 1939. Development and Perspectives of Landscape Ecology 2002 ed.

Turban, E., Rainer, R.K. and Potter, R.E. 2005. *Introduction to Information Technology*. Hoboken, NJ: John Wiley & Sons.

Turky, Ayad Mashaan, Mohd Sharifuddin Ahmad, Mohd Zaliman, and Mohd Yusoff. 2009. The Use of Genetic Algorhythm for Traffic Light and Pedestrain Crossing Control. *IJCSNS International Journal of Computer Science and Network Security* 9 (2): 88–96.

UN DESA'S Population Division. 2015. World's Population Increasingly Urban with More than Half Living in Urban Areas. United Nations. http://www.un.org/en/development/desa/news/population/world-urbanization-prospects-2014.html.

UN World Water Development Report. 2017. *Wastewater, the Untapped Resource*. World Water Assessment Programme (WWAP). Paris: United Nations Educational, Scientific and Cultural Organization.

UNCTAD. 2012. *Geospatial Science and Technology for Development with a Focus on Urban Development, Land Administration and Disaster Risk Management. UNCTAD/DTL/STICT/2012/3*. New York and Geneva: United Nations.

UN-Habitat: *The Challenge of Slums Global report on Human Settlements*. London and Sterling, VA: Earthscan; 2006.

Unhelkar, Bhuvan. 2011. *Handbook of Research on Green ICT*. Hershey, PA: Information Science Reference.

United Nations. n.d. Sustainable Development Goals (SDGs) 8, 9, and 10. United Nations Department of Public Information. https://sustainabledevelopment.un.org/?menu=1300.

United Nations Department of Economic and Social Affairs (DESA). 2015. *The Critical Role of Water in Achieving the Sustainable Development Goals: Synthesis of Knowledge and Recommendations for Effective Framing, Monitoring, and Capacity Development*. New York: United Nations.

United Nations Department of Economic and Social Affairs/Population Division. 2012. *World Urbanization Prospects*. New York: United Nations.

United Nations Economic and Social Commission for Asia and the Pacific (ESCAP) and Regional Integrated Multi-Hazard Early Warning System (RIMES). 2016. *Flood Forecasting and Early Warning in Transboundary River Basins: A Toolkit*. Bangkok: ESCAP.

United Nations Educational Scientific and Cultural Organization. 2009. Water in a Changing World. http://unesdoc.unesco.org/images/0018/001819/18199 3e.pdf

United Nations. 2013.SDG 7. *Goal 7: Ensure Access to Affordable, Reliable, Sustainable and Modern Energy for All*. New York: United Nations Statistics Division Statistical Services Branch.

United Nations, Department of Economic and Social Affairs, Population Division. 2017. *World Population Prospects: The 2017 Revision*. New York: United Nations.

United Nations Economic and Social Council, 2013. *Science, Technology and Innovation for Sustainable Cities and Peri-urban Communities*. Commission on Science and Technology for Development. Geneva: United Nations.

United States Department of Agriculture. 2002. Nanoscale science and engineering for agriculture and food systems; Report submitted to Cooperative State Research, Education and Extension Service, United States Department of Agriculture, National Planning Workshop; November 18–19; Washington, DC.

U.S. Army Composite Risk Management (CRM). 2017. Army Knowledge Online, Headquarters, Department of the Army. Chapter 3.

U.S. Energy Information Administration. 2017. *International Energy Outlook* 2017. Washington, DC: U.S. Energy Information Administration.

Vandenbroucke, Danny. 2017. Urban Planning and Smart Cities: When Is 'Smart' Really Smart. Geosmartcities Conference, Genoa February 15.

Vasiljevic-Shikaleska, Aneta, Biljana Gjozinska, and Martin Stojanovikj. 2017. The Circular Economy – A Pathway to Sustainable Future. *Journal of Sustainable Development* 7 (17): 13–30.

Venton, P. and S. La Trobe. 2008. *Linking Climate Change Adaptation and Disaster Risk Reduction*. Teddington, UK: Tearfund.

Vickery, G. 2012. Smarter and Greener? Information Technology and the Environment: Positive or Negative Impacts? Winnipeg, Canada: The International Institute for Sustainable Development.

Vigneshvar, S., C.C. Sudhakumari, B. Senthilkumaran, H. Prakash. 2016. Recent Advances in Biosensor Technology for Potential Applications – An Overview. *Frontiers in Bioengineering and Biotechnology* 4: 11. doi:10.3389/fbioe.2016.00011.

Vine, Edward. 2012. Adaptation of California's Electricity Sector to Climate Change. Climatic Change 111: 75–99. 10.1007/s10584-011-0242-2.

Visa, Ion. 2014. Sustainable Energy in the Built Environment – Steps towards NZEB. In *Proceedings of the Conference for Sustainable Energy (CSE) 2014*. Cham: Springer.

Visser, W. and G. Brundtland. 2009. *Our Common Future*. n.p.: The Brundtland Report: World Commission on Environment and Development.

Walter Castelnovo, Gianluca Misuraca, and Alberto Savoldelli. 2015. Smart Cities Governance: The Need for a Holistic Approach to Assessing Urban Participatory Policy Making. *Social Science Computer Review* 34. 10.1177/0894439315611103.

Wan, Yang, and Yudong Yao. 2001. Measuring Economic Downside Risk and Severity. Washington, DC: World Bank.

Wang, Jianqiang, Lei Zhang, and Keqiang Li. 2012. An Adaptive Longitudinal Driving Assistance System Based on Driver Characteristics. *IEEE Transactions on Intelligent Transportation Systems* 14 (1): 1–12.

Wang, Zhong Lin, and Wenzhuo Wu. 2012. Nanotechnology-Enabled Energy Harvesting for SelfPowered Micro-/Nanosystems. *Angewandte Reviews*. doi:10.1002/anie.201201656.

West-Eberhard, M.J. 2003. *Developmental Plasticity and Evolution*. New York: Oxford University Press.

Wilkinson, P.J., and D.G. Cole. 2010. The Role of Radio Science in Disaster Management. The Radio Science Bulletin. IPS, Bureau of Meteorology, Australia, no. 335.

Williams, R. H., E. D. Larson, and M. H. Ross. 1987. Materials, Affluence, and Industrial Use. *Annual Review of Energy* 12: 99–144.

William, J. Kramer, Beth Jenkins, and Robert S. Katz. 2007. The Role of Information and Communication Technologies. http://www.hks.harvard.edu/mrcbg/CSRI/publications/report_22_EO%20ICT%20Final.pdf [Accessed 12 September 2012].

Wimo A., L. Jonsson, and B. Winblad. 2006. An Estimate of the Worldwide Prevalence and Direct Costs of Dementia in 2003. *Dementia and Geriatric Cognitive Disorders* 21: 175–81.

Winograd, M. and A. Farrow. 2011. Sustainable Development Indicators for Decision Making: Concepts, Methods, Definitions and Use. Encyclopedia of life support system (ecolss). http://www.eolss.net/ebooks/Sample%20Chapters/C13/E1-46B-02.pdf

Winston Maxwell, and Marc Bourreau. 2014. Technology Neutrality in Internet, Telecoms and Data Protection Regulation. Hogan Lovells Global Media and Communications. https://www.hoganlovells.com/~/media/hogan-lovells/pdf/publication/2015_21_ctlr_issue_1_maxwell.pdf?la=en.

Wootton, Richard, John Craig, and Victor Patterson. 2006. *Introduction to Telemedicine*. London: Royal Society of Medicine Press.

World Bank. 2012. *Improving the Assessment of Disaster Risks to Strengthen Financial Resilience*. Washington, DC: World Bank.

World Economic and Social Survey 2017: Reflecting on Seventy Years of Development Policy Analysis. 2017. New York: United Nations.

World Health Organization. 1996. *Climate Change and Human Health: Risks and Responses*. Geneva: World Health Organization.

World Health Organization. 2004. *Safe Health-Care Waste Management*. Geneva: World Health Organization.

World Health Organization. 2005. *International Health Regulations*, 3rd Ed. Geneva: World Health Organization.

World Health Organization. 2008. *Building Foundations for eHealth in Europe: Report of the WHO Global Observatory for eHealth*. Geneva: World Health Organization.

World Health Organization. 2018. Drinking-Water: Key Facts. Factsheet on sanitation Water Sustainability hrough Nanotechnology: Nanoscale Solutions for a Global-Scale Challenge. 2016. https://www.nano.gov/sites/default/files/pub_resource/water-nanotechnology-signature-initiative-whitepaper-final.pdf.

World Water Assessment Programme. 2009. The United Nations World Water Development Report 3: Water in a Changing World. Paris: UNESCO, and London: Earthscan.

WWF Sweden. 2008. The Potential Global CO_2 Reductions from ICT. Identifying and Assessing the Opportunities to Reduce the First Billion Tonnes of CO_2. Solna, Sweden: WWF.

Yellowlees P.M., K. Chorba, M.B. Parish, H. Wynn-Jones, and N. Nafiz. 2010. Telemedicine Can Make Healthcare Greener. *Telemedicine Journal and e-Health* 16: 230–33.

Yermack, David. 2017. Corporate Governance and Blockchains, *Review of Finance* 21 (1): 7–31. https://doi.org/10.1093/rof/rfw074

Younger M., H.R. Morrow-Almeida, S.M. Vindigni, and A.L. Dannenberg. 2008. The Built Environment, Climate Change, and Health: Opportunities for Co-benefits. American Journal of Preventive Medicine 35: 517–26.

Yuka Maesawa, Yoshihiki Hatakeyama, Mutsumi Saito, and Fukutaro Hirota. 2014. Conservation of Biodiversity by Making Use of ICT. *Fujitsu Scientific & Technical Journal* 50 (4): 44–51.

Zhang, Hao, Gitesh Raikundalia, Yanchun Zhang, and Xinghuo Yu. 2009. Application of Multi-agent Technology to Information Systems: An Agent-based Design Architecture for Decision Support Systems. *Australian Journal of Information Systems.* 15. 10.3127/ajis.v15i2.24.

Zhou Guangwei. 2006. Optimization of Adaptive Traffic Signal Control with Transit Signal Priority at Isolated Intersections Using Parallel Genetic Algorithms. *ProQuest ETD Collection for FIU.* AAI3217586.

Index